Digital Filter Design using Python for Power Engineering Applications

Shivkumar Venkatraman Iyer

Digital Filter Design using Python for Power Engineering Applications

An Open Source Guide

 Springer

Shivkumar Venkatraman Iyer
Hamburg, Germany

ISBN 978-3-030-61862-9   ISBN 978-3-030-61860-5   (eBook)
https://doi.org/10.1007/978-3-030-61860-5

This Springer imprint is published by the registered company Springer Nature Switzerland AG
The registered company address is: Gewerbestrasse 11, 6330 Cham, Switzerland

*To the everlasting memories
of my dear departed friend
and mentor Mukundan Raghavan.*

Preface

The biggest transformation in the power engineering sector has been the pace at which it has been digitized. This trend will continue in the years to come, where traditional analog devices are being replaced by computerized devices. The advent of smart grids makes the implementation of digital technology its core philosophy. This transformation is in general a positive change since we are moving towards a power system that is better coordinated and more efficient allowing a greater proliferation of green technologies. The transformation also poses a challenge to the traditional power system design and planning methodology, where such digitization was not present. The power engineer of today needs to be familiar with digital technologies in addition to power engineering.

As a power engineer, I have been fortunate to have had the opportunities to learn digital technologies while the transformation of the power system was still in the nascent stage. The prospect is far more daunting now. Due to these hurdles, it will be essential that digital skills in power engineering are presented with a unique perspective. The current state of digital education for power engineers requires them to enroll for courses that are taught to students in telecommunications. However, this approach to education does not provide a power engineer with the insight into how to apply digital technologies in power systems. It is left to the students to figure out how to apply this knowledge in the power systems domain.

As a power engineer, most of the solutions that I implemented for digital control or digital signal processing in power electronics applications were heuristic. Unfortunately, the lack of adequate training and pressing deadlines did not leave much time to formulate an analytical approach. However, this approach of trial-and-error results in sub-optimal solutions that over time can jeopardize the operation of systems. Gradually, over time, I went back to theory and redesigned my solutions using an analytical approach. The result was that the new solutions were repeatable and scalable and the guesswork involved was reduced to a bare minimum that would be present in any design activity. Over the past few years, I have begun to share my knowledge through online courses and books, as I have found that access to educational content in power engineering is still limited.

The objective of this book will be to present digital technologies such as signal processing and control systems to power engineers from a practical perspective. Any book on signal processing or control systems will not be complete without dealing with theory in detail. However, the presentation of theory will be keeping in mind that the reader is a power engineer who needs to implement a solution in a hardware prototype. This is the unique selling point of this book. It is a book written by a power engineer and for power engineers, specifically taking into account that the traditional treatment of mathematical theory will not be useful for power engineering applications.

This book will not be considered "heavy" reading. This is intentional, as the target audience of these books are not just students of power engineering but also practicing electrical engineers. The challenges in continuing education faced by practicing professionals are fairly acute, as most courses either do not take into account the limitation of time and energy that can be devoted towards learning a topic or else are very superficial and aim to provide an overview of a topic rather than an in-depth dissection. To ensure fair participation of mature students, my books use analogies, simplified equations, block diagrams and a detailed narrative as opposed to a heavily mathematical approach. To provide an in-depth understanding of how to implement solutions, programming and simulation are provided at every stage.

This book uses only free and open source software, most notably the Python programming language. As an open source enthusiast, I am an ardent proponent of open source technology and use exclusively open source tools for all my programming requirements. The use of Python programming language is to encourage the use of open source tools in core engineering, which currently is heavily proliferated by proprietary enterprise software. Besides the affordability aspect of open source tools, open source technology is community driven, which eventually provides freedom and choice to the end user. Online forums feature active discussions on many topics and provide assistance to users besides allowing for bug fixes and software updates.

To sum up, this book is about discovering a practical approach to applying signal processing in power engineering applications by using open source software. The solutions provided have been generalized to the greatest possible extent such that an engineer can use the approach for designing customized solutions for any given application. The use of open source tools guarantees the availability of these solutions free of licensing requirements. The eventual aim is to demystify mathematical topics such as signal processing and control systems by emphasizing on the final practical implementation while ensuring a strong link between theory and practice.

Acknowledgements

I would like to thank Ramdas Murdeshwar for his continuous support as I developed my online courses and wrote this book. I would like to thank Janamejaya C for being a loyal follower of the project and his constant encouragement. I would like to thank Krishnan Srinivasarengan for his support and encouragement. I would like to thank Prof. Bin Wu for being a supportive mentor as I embarked on becoming an author of technical books. In addition, my greatest thanks is reserved for Rakhat for her constant support, encouragement and patience as I spent countless hours on this book project.

Hamburg, Germany   Shivkumar Venkatraman Iyer

Contents

1 Introduction 1
  1.1 The Importance of Digital Signal Processing 1
  1.2 Why Another Book on DSP? 2
  1.3 The Use of Computational Tools 4
  1.4 Philosophy Behind the Book 5
  1.5 Outline of the Book 6
  1.6 How to Read This Book 8

2 Introduction to Discrete Systems 9
  2.1 Introduction 9
  2.2 Discrete Versus Continuous 10
  2.3 Continuous Time Signal Processing Example 11
  2.4 The Need for Digital Signal Processing 12
  2.5 Digital Signal Processing 13
  2.6 Advantages of Digital Signal Processing 15
  2.7 Process of Converting a Signal to Digital Form 16
  2.8 Practical Analog to Digital Conversion 18
  2.9 How Would a Processor Handle the ADC? 19
  2.10 Conclusions 20
  References 21

3 Getting Started with Signal Processing 23
  3.1 Introduction 23
  3.2 Reviewing Capacitors and Inductors 23
  3.3 Filters as Combinations of Inductors and Capacitors 25
  3.4 The Concept of Transformations 27
  3.5 Laplace Transform 28
  3.6 Revisiting Inductors and Capacitors with Laplace Transform 31
  3.7 What About the Original Variables? 34

3.8   The Mystery Behind s   35
3.9   Advantage of Laplace Transform   41
3.10   Digital Signal Processing..   43
3.11   Continuous to Digital Conversion   48
3.12   Putting the Pieces Together   51
3.13   Conclusions   52
References   52

4   **Introduction to Python**   53
4.1   Introduction   53
4.2   Basic Programming Setup..   54
4.3   Generating Signals: Creating Numpy Arrays   55
4.4   Array Manipulation..   57
4.5   Generating Waveforms: Sine and Cosine Functions   58
4.6   Plotting with Matplotlib   60
4.7   Plotting Waveforms..   64
4.8   Conclusions   69
References   69

5   **Implementing Analog Filters Digitally**   71
5.1   Introduction   71
5.2   Implementing a Capacitor Filter Digitally...........................   72
5.3   Coding the Capacitor Filter   74
5.4   The Inductor Filter...   82
5.5   Coding the Inductor Filter...   84
5.6   The dc Offset..   88
5.7   Do We Have Offsets in Reality?   90
5.8   Lossy Capacitor..   92
5.9   Mathematical Model of a Lossy Inductor   98
5.10   Modeling the LC Filter   102
5.11   Analysing the Performance of the LC Filter   106
5.12   Conclusions   110
References   111

6   **Frequency Response Analysis**   113
6.1   Introduction   113
6.2   A New Look at Complex Numbers   114
6.3   Transfer Functions: Magnitude and Phase Angle   115
6.4   Transfer Function: Frequency Response   119
6.5   Bode Plots..   121
6.6   Getting Started with Scipy.Signal....................................   124
6.7   Frequency Response of an Inductor   127
6.8   Interpreting the Inductor Frequency Response   130
6.9   Frequency Response of an LC Filter.................................   134
6.10   Physical Significance of these Frequency Responses   138

	6.11	Generalized Poles and Zeros	140
		6.11.1 The First Order Pole	140
		6.11.2 The Generalized Second Order Pole	143
		6.11.3 Generalized First Order Zero	144
		6.11.4 Generalized Second Order Zero	145
	6.12	Conclusions	147
	References		147
7	**Filter Design**		**149**
	7.1	Introduction	149
	7.2	Designing Filters for Power Applications	150
	7.3	Getting Started with Low Pass Filter Design	151
	7.4	Simulating a Filter Transfer Function	157
	7.5	Cascading Transfer Functions	162
	7.6	Simulating the Cascaded Low Pass Filter	166
	7.7	Getting Started with a Notch Filter	170
	7.8	Adding Zeros to the Transfer Function	172
	7.9	Normalizing the Transfer Function	176
	7.10	Conclusions	184
	References		185
8	**Conclusions**		**187**
	8.1	A Summary of the Course Contents	187
	8.2	A Few Comments on the Approach	189
	8.3	Scope for Future Work	190
Index			191

Chapter 1
Introduction

1.1 The Importance of Digital Signal Processing

The power system has rapidly changed over the past few decades. Along with an augmentation of infrastructure to support economic growth and development, the power system has been heavily digitized at every level of the transmission and distribution stages. With the rapid move towards making systems "smarter", the goal is to use power system measurements along with communications to control and coordinate segments of the power system to operate more efficiently and with greater reliability. As an example, conventional relays are being replaced by numerical relays, while traditional electromagnetic sensors are being replaced by Phasor Measurement Units (PMUs)—devices that have on-board microcontrollers, radio frequency measurement, high capacity storage and connectivity to central network operators.

When I was an undergraduate student of electrical engineering, learning digital techniques was primarily to augment your knowledge and to have an insight into techniques used in communications and other digital fields. In the late 1990s, very few electrical engineers thought that the power system would be flooded with digital technology as it is today. In modern times, electrical engineers and for that matter any engineer has accepted the fact that whatever maybe their field, it will soon be digitized and will even have a significant penetration of technologies such as artificial intelligence and machine learning. Nowadays, your car is as smart or ever smarter than your home computer. The exposure that core engineers have to digital technologies is far greater than a few decades back.

Despite the need to learn digital technologies, there is a lack of educational resources to bridge the gap between conventional core engineering such as power systems and the latest digital techniques such as digital signal processing (DSP). The digital technologies are taught from the perspective of communications and other pure Information Technology (IT) fields. Eventually, the power engineer has to

S. V. Iyer, *Digital Filter Design using Python for Power Engineering Applications*,
https://doi.org/10.1007/978-3-030-61860-5_1

figure out how to modify the application for the power system as almost no concept can be transplanted from one domain to another without any changes. Filtering techniques used in voice communications will be very different as compared to the need for filtering in a power system.

In order that the changing power system remains stable and reliable, it is essential that those designing, operating and maintaining it are comfortable with the digital technologies penetrating it. If these digital technologies are perceived as black boxes such that if they fail, the only solution is to "Call IT services", the power system reliability is at risk. Besides giving independence to the power system operators, the benefit of learning digital technologies for power engineers is to enable the conversion of analog solutions to digital form. With the falling price of microcontrollers and wireless communication interfaces, conventional control and processing implemented in power systems can be implemented digitally instead.

How would a digital implementation of well-tested hardware help? Digital implementation offers far greater flexibility as opposed to an electromechanical implementation. For example, a numerical relay based on microcontrollers can issue a command to a circuit breaker to disconnect a circuit in exactly the same way as an electromechanical relay would. However, in the case of a numerical relay that communicates with a central operator, the settings of the numerical relay can be altered remotely causing it to behave with a different level of sensitivity. On the other hand, the sensitivity of an electromechanical relay is based on its physical construction, which is very difficult to alter. In modern power systems with a large degree of variability, flexibility of operation adds greater reliability to the power system. As an example, the large penetration of renewable energy technology such as wind farms could need rapid deployment of voltage stabilizers to prevent voltage fluctuations at load centres.

1.2 Why Another Book on DSP?

Signal processing and the digital counterpart DSP have been studied for around two centuries with some of the earlier works published during the Napoleonic era of the early 1800s. The last few decades have seen an explosion of international journals, periodicals and conferences where researchers have raced to publish their work. Therefore, if an engineer wished to study DSP, the material available is vast. And there is where the problem occurs. An engineer might be at a loss of where to begin unless the course curriculum at the university where he or she studies is explicit in the recommended texts to be followed.

This problem is worse for power engineers. As a student, I read several books on signal processing and though many of them were excellent references to learn the basics of signal processing theory, none of them had any references to power engineering. Reading a book on signal processing with the intention of implement-ing the concepts in power electronics which was my chosen specialization left me clueless on how I should take the first steps. The books were written for purely digital applications such as communications, computer vision, or a similar topic. How was I to take those concepts and use them for implementing a filter using a microcontroller for a power electronics application?

During my years as a graduate student followed by stints in various power companies, I was assigned several projects that needed controls and filters implemented in digital platforms such as microcontrollers and Field Programmable Gate Arrays (FPGAs). After a while, I formulated a few design guidelines that were a combination of formulas scraped off some signal processing books and a few observations I made while implementing hardware. Most of these were heuristic and needed fine-tuning by trial and error methods. Eventually, when they did work, it was customary to leave a comment—"DO NOT MODIFY THIS" as redesigning would be another agonizing process. I had the privilege of working with many skilled and talented power engineers and found to my surprise that their approach to digital implementation of filters was similar to mine.

Over the past few years, I have begun creating educational material for power engineers. My motivation was that this helpless groping in the dark until a solution somehow appears within grasp that is prevalent in power engineering needs to change. As the digitization of the power system accelerates, well-designed solutions are absolutely essential as opposed to heuristic solutions. An abundance of digital implementation generated through heuristic means poses an immense reliability risk to the power system. For this purpose, my philosophy in creating educational material whether through books or through online courses has been to the reverse of the traditional approach. I decided to approach signal processing primarily from the viewpoint of the final application.

In this book on DSP, I tackle the problem on how to design filters in the digital domain for implementation with platforms such as microcontrollers. This book is not a comprehensive reference to digital signal processing. This book talks about how to implement digital filters. That is it. We will present methods to design, analyze and verify filters from the first principle using a combination of basic signal processing theory and computational tools. We will aim to establish a well-defined approach to arriving at a solution stressing on procedures that can be replicated, reproduced and changed over and over again. The main objective is—the design of a digital filter should not be a magical heuristic process but a step-by-step analytical procedure.

It is impossible to describe the implementation of a digital filter without a background on basic signal processing. In this book, I will cover basic signal processing theory. However, theory will not be covered from a purely mathematical viewpoint, but theory will be broken down to core engineering concepts that a power engineer is well-versed and comfortable with. Throughout the book, we will cross the bridge between the mathematics behind signal processing and the engineering applications that a power engineer is finally interested in.

In this approach, to help solve the practical problem of designing digital filters, I believe this book is unique. This book is written by a power engineer who has felt the pain in trying to implement mathematical theory in core engineering. I have written every chapter of this book with that in mind—that the final readers are power engineers who need to get their hardware to work. In terms of mathematics, this book covers only a fraction of what any traditional signal processing book will cover. In terms of mathematical rigour, this book may appear weak and at sometimes even misleading because this is not a book of mathematics, but instead a book of applying mathematics to engineering. Theory has been expressed in ways that are understandable to an engineer rather than for its mathematical authenticity.

1.3   The Use of Computational Tools

Every book of signal processing contains examples using a scientific software. MATLAB is very commonly used in most signal processing books as it has an extensive signal processing blockset. The use of MATLAB for a book on signal processing is also a natural choice since most universities in the world maintain academic licenses for MATLAB. Besides MATLAB, Scilab is another free and open source alternative also with an extensive signal processing library. One of the drawbacks of MATLAB is that it is a fairly expensive software that might make it out of reach of smaller universities particularly in developing countries.

As a software developer and a Linux enthusiast, I have always taken a keen interest in open source software. Open source software is community driven, developed and maintained by a combination of paid developers and volunteers. Many open source software begin as side projects undertaken by developers, which are then managed by a team of volunteers. As the open source projects grow and begin to be used by companies for commercial projects, many projects begin to receive sponsorship from corporations. This enables open source projects to recruit a core team of paid developers to ensure stability and continuous growth of the project. However, the open source nature of projects invariably results in these projects receiving a significant amount of community support from volunteer developers whether it be in bug reporting or fixes.

One of the most popular open source projects in the scientific computing arena is Python. Created in the 1990s by Guido van Rossum, the free and open source software has been adopted by almost every area of computing—general programming, web development, data analytics, graphical user interface development and artificial intelligence. Python has a vibrant and active developer community with continuous development and extensive documentation. Most importantly, Python has set itself as an alternative to MATLAB and Scilab in the recent years.

In this book, I will use Python and associated packages developed using Python for all the examples. NumPy (Numeric Python) forms the backbone of scientific computing using Python as it provides a vast collection of matrix computation functions. SciPy (Scientific Python) builds on the NumPy library with an array of functions for signal processing. The book will contain several code examples along with description of the functions being used.

The book will start with the basics of Python programming with a description of some of the basic Python commands. Though the book is not a comprehensive

Python reference, the introduction of basic Python programming will help an engineer new to Python to understand the more advanced functions used while designing filters. The book will gradually introduce Python and NumPy commands by describing how to generate signals. The book uses MatPlotLib, which is a plotting library developed using Python to generate all waveforms and plots. The book will simulate digital implementation of analog filters to begin introducing the concept of digital filtering.

Using SciPy, we will examine how to express transfer functions and generate frequency response characteristics to examine the behaviour of filters for a wide range of frequencies. We will examine how we can use SciPy to express filters in the digital domain and implement them in simulations. We will also examine functions to synthesize higher order transfer functions. The code examples will describe how the design of the filter can be converted into a well-structured and easily modifiable process. The process can be undertaken repeatedly with minimal effort and all we have to do is modify parameters.

The book will present two filter design cases that will be covered in-depth. The filters will be designed incrementally using a pattern of design, analysis and verification. We will use analytical tools to design filters, judge their performance and verify our judgement with simulations. The examples will contain extensive code samples that will focus on the intricacies of the Python commands being used, potential hurdles and ways that we can overcome them.

1.4 Philosophy Behind the Book

As an author writing about topics that might seem highly saturated, it is important for the reader to be made aware of the philosophy behind this book and also subsequent books in the future. In these modern times, technology changes rapidly and what we might learn today may be obsolete in another ten years. Therefore, whether we are students or practicing engineers, it is absolutely essential to continuously upgrade our knowledge to stay in touch with changes in technology. I would say that I have learned more while working in industry than while I was a student for the simple reason that knowledge gained while working towards concrete goals is far more precious than knowledge gained as a part of completing a course.

As an online educator and in the process of writing this book, I found myself learning all over again. To be able to put together this book and explain concepts gleaned from a vast number of sources, I spent a considerable amount of time reading on basic mathematical theory. I would urge the readers of this book and for that matter any serious engineer to make the process of learning a continuous one. The assumption that without formal training, knowledge gained is merely "coffee talk" is totally and completely false. True knowledge is that which is applied and continuously improved upon irrespective of how it has been acquired.

It is unfortunate that even in these modern times, many aspiring engineers do not have access to resources to continuously update their knowledge. On a positive note, many of the best universities in the world have made their course materials openly accessible so that students all over the world can benefit. However, most of the material is still presented in a manner that it can be followed by a reader who is a full-time student at a university. For a practicing engineer in industry, many of these open access course material are still of limited use. There is still a need to generate educational material for universal use that can be followed by students, working engineers and also mature students.

This book is an attempt to reach out to everyone who is interested in digital signal processing. This book is as self-contained as possible with every concept explained in as simple terms as possible. There may even be times when explanations of mathematical concepts might have been watered down to an extent that they might be almost incorrect. However, the purpose behind this presentation is to make theory accessible to everyone irrespective of their background. While writing this book, I have used the Feynmann theory of teaching. To begin with, choose a topic that I would like to teach even if it is a topic that I am not an expert in. Next, I attempt to describe the topic to a child with little or no previous knowledge of the topic. During this process of explanation, I examine gaps in my own knowledge, which have driven me back to gathering more information. I return to improve the explanation until the topic is universally understandable.

1.5 Outline of the Book

The book consists of six chapters excluding the introduction and conclusion. The contents of the chapters are as follows.

Chapter 2 introduces the concept of a digital or discrete system. We ease into the discussion of a digital system with some non-engineering examples that everyone can relate to. We will then examine the meaning of a digital system from the engineering perspective while examining specific examples related to electrical engineering. Before describing the process of expressing a system in digital form, we will examine some of the advantages of digital systems over analog systems. We will then examine how sampling and quantization are the defining factors in any digital system and will briefly describe these processes with some examples. We will describe the process of analog to digital conversion and describe how the process of conversion takes place in hardware.

Chapter 3 introduces basic signal processing theory. We examine using basic network equations how passive components such as inductors and capacitors behave like filters in analog circuits. We introduce the Laplace Transform and examine the effect it has on circuits by examining how the equations of passive components are transformed. We will examine the advantages of Laplace Transform and the reason why it has become so popular in signal processing. We will then meander through some scattered mathematical concepts to describe how the Laplace Transform enables us to perform frequency response analysis that is the cornerstone of signal processing. We will examine the difference between frequency response analysis in the continuous domain and the digital domain. We will examine how we can perceive frequency as an operator that will enable us to implement a digital filter.

Chapter 4 is a brief description of some of the basic Python commands, which will be used in the later chapters. This chapter will examine some of the commands used for generating signals, plotting them and manipulating them. We will examine some basic coding examples to generate waveforms using MatPlotLib. This chapter is not a comprehensive tutorial on Python but is merely to make the book more self-contained. If the reader wishes an in-depth tutorial on basic Python, there are many free and paid resources online.

Chapter 5 will examine how analog filters can be expressed digitally. The purpose of this chapter is to provide a bridge for an electrical engineer who is comfortable with analog circuits so as to ease into the process of digital signal processing. We will examine how analog circuits can be transformed into a frequency domain representation using Laplace Transform. Subsequently, the frequency domain representation will be converted from the continuous domain to the digital domain. The entire process will be performed manually by writing equations and solving them so as to bridge the gap between mathematical equations and programmatic simulations. The chapter will also describe some of the practical issues related to digital implementation and how eventually analog circuits can be digitally represented with a fair degree of accuracy.

Chapter 6 will introduce the concept of frequency response analysis. We will examine how analog circuits do not behave in the same manner for all frequencies. We will examine how performing Laplace Transform on time domain equations results in frequency domain equations that allow us to study the frequency dependent behaviour of analog circuits. We will examine the Python commands that can be used to represent a transfer function as well as the commands with which the frequency response characteristics can be generated and plotted. We will examine how transfer functions can be interpreted and their behaviour translated to frequency response characteristics. We will examine the transfer functions of some basic analog circuits and interpret the behaviour of the circuits from the frequency response characteristics. We will then examine how we can use a number of transfer functions as basic building blocks with their frequency response characteristics being well defined.

Chapter 7 will combine all our learning from the previous chapters to design filters. We will design a low pass filter and a band stop filter. We will use transfer functions from Chap. 6 to synthesize our filters. We will introduce several Python commands that will automate the process of synthesizing filters, analyzing their frequency response characteristics and implementing them in a simulation by expressing them in the digital domain. The chapter will describe the process of how to design a filter—to begin with an initial hypothesis, examine the result using frequency response characteristics and improve on the design by synthesizing higher order transfer functions. We will examine how a design fulfills the requirements after it passes every requirement laid down in the initial specification. We will simulate each filter to verify our final designs. This chapter will heavily use programming and will use the minimal amount of manual calculations to describe how filter design can be automatized.

1.6   How to Read This Book

Reading any book from cover to cover needs a fair degree of dedication. Before reading this book, it is important for the reader to keep in mind that this is not a comprehensive book on digital signal processing. The objective of this book is to present a simplified procedure towards digital filter design. The signal processing theory covered in the book is to help make the examples in the final chapter understandable.

For a reader to make the most of this book, it is important that the reader must be willing to write code. The later chapters will contain several code samples with detailed description on the working of the code. The book contains no theoretical exercises, no numerical problems and no quizzes. The book describes the process of filter design combining theory and programming. The main substance of the book is in Chap. 7 where we will design two sample filters. The chapters before that are only for the reader to be able to appreciate the filter design process.

The only advice I would offer to the reader of this book is—be prepared to code. Install all the necessary software and code along as you read the book and whenever you find code samples. The code presented in this book is also available as downloadable resources, so feel free to tear them down. Most importantly, try to use the concepts learned in this book to design filters for your own projects.

Chapter 2
Introduction to Discrete Systems

2.1 Introduction

In this chapter, we will get started with digital signal processing at a very high level. We will not get started with any theory or implementation as those will follow in later chapters. In this chapter, the reader will get acquainted with the basic idea of how digital signal processing is achieved.

To begin with, I will use very basic examples to talk about discrete systems and how this concept is something we use every day without even knowing it. I will talk about the need to perform signal processing and what are some common examples of signal processing such as filtering that is already found on circuits that we use every day. We will gradually look at how digital signal processing can overcome some of the disadvantages of analog or continuous time signal processing and some cases where the advantages are fairly obvious.

We will then gradually formalize what is a discrete system by introducing the concepts of sampling and quantization. We will examine using a very simple example of room temperature measurement, how a measured quantity can be expressed in digital form. We will then introduce the concept of analog to digital conversions, once again at a high level. We will finally look at how a processor can communicate with an analog to digital converter.

This chapter serves the purpose of easing you into the later chapters where I use terms like sampling in greater detail. The objective is to make it easier to understand theoretical concepts by first understanding how digital signal processing is achieved in practice.

S. V. Iyer, *Digital Filter Design using Python for Power Engineering Applications*,
https://doi.org/10.1007/978-3-030-61860-5_2

2.2 Discrete Versus Continuous

How and where did we start with the concept of discrete? Without getting too much into the history of when it started and who was the first pioneer in this field, let us try to understand this concept in the real-world scenario.

In the real world, almost every process is continuous. You live continuously. There are no periods for which you do not exist. When you watch a soccer match, the match is a continuous process. There are 22 players on the field chasing the football all the time. Technically, there are periods for which the match stops—free kicks, player substitution, changing sides, etc. But there is no such time period for which the game does not exist. Something is happening. If a player is injured, something is still happening—he is lying on the ground, in pain, some players are protesting, saying he is faking it, others are saying the offending player should be given at least a yellow card, and the referee wants the player off the field as soon as possible to minimize time wasted. So, really, there is something continuously happening and there is always entertainment particularly if you take the commentary into account since that almost never stops.

So, the match can be said to be a continuous process. Now let us look at an example of a discrete process. In the same context of soccer, a few different simple examples of discrete systems are the viewers of this match. Of course, depending on how passionate someone is about soccer, there would be different types of viewers. There would be those who would be glued to the TV screen the entire time, not moving, not taking their eyes of the screen until the final whistle goes. There would be those who come and go, busy with their work, changing channels and watching something else, or just busy with their household chores and just checking in once in a while in case the game seems to be getting interesting.

Let us take the example of the passionate fan who cannot take his or her eyes off the screen until the match ends. He or she sees everything that happens, including the parts that you might call useless—player injured, substitutions and everything. He or she follows commentary and analysis, posting on social media as the match progresses. The type knows every player on the field and every bit of information about soccer in general. In this case, the process of observation of the game by this viewer can also be said to be continuous. He or she sees everything, reacts to almost everything and knows everything. Almost every one of us knows at least one such sports lover.

Now let us take the example of the casual viewer. He or she watches parts of the match every now and then. He or she might miss some useless events like injury time and also miss major events like a goal if he or she was watching another channel or engaged in some other work. And a very large amount of viewers will fall in this category as most people just do not have the time to follow a match completely from beginning to end. In this case, the observation of the match by this viewer can be termed as discrete, which by its definition means "individually separate and distinct" [1, 2].

The same can be said about signal processing in the real world. When you wish to process signals measured from a physical system, you can do so continuously or discretely. And just like with the two soccer viewers, the effort needed to perform continuous versus discrete processing is different. In one case, we must be capturing data continuously without any break, and in the other, we can afford to collect data once in a while.

2.3   Continuous Time Signal Processing Example

Now that we have described the difference between discrete and continuous processes in the non-electrical sense, let us look at an electrical example of continuous signal processing.

Suppose we are measuring the voltage across two nodes in a circuit. This measured voltage is being stepped down, in many cases isolated using a Hall effect sensor [3], conditioned using analog circuits and is finally fed to a control circuit. A very typical example of signal processing that is done using analog circuits is to remove high frequency noise from this measurement before feeding it to the control circuit. This is common because you do not want our control to be affected by measurement noise.

The solution is quite simply to connect a filter capacitor across the output of the Hall effect sensor as shown in Fig. 2.1. The capacitor by its nature acts as a sink of high frequency signals because the reactance of a capacitor decreases as frequency of a signal increases:

$$X_c = \frac{1}{2\pi f C} \tag{2.1}$$

Moreover, the capacitor provides a smoothing effect with respect to the voltage across it because the capacitor opposes rapid fluctuation of the voltage across it:

$$v = \frac{1}{C} \int i \, dt \tag{2.2}$$

Such a circuit is an example of a continuous low pass filter. This is because the capacitor is physically connected and will play its role as long as a measured signal is fed to it.

Fig. 2.1 An example of a filter capacitor for a voltage sensor

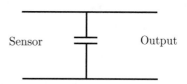

And this kind of filter is very popular and is commonly used even in advanced control. Why? Capacitors are cheap and do not occupy much space. Take a look at the printed circuit board (PCB) of any commercial electronic appliance. It will be full of capacitors. Most capacitors used in these circuits have parametric values ranging from a new picofarad to a few nanofarads and a voltage rating of up to 35 V. Such capacitors would cost a few cents and even less if purchased in bulk. Even if more advanced noise removal is performed inside the controller, this first level of filtering is very convenient to implement. And capacitors do not occupy a lot of space, which is why a PCB could have hundreds of them, and the increase in size would be a few square centimetres at the most.

The downside to such filtering is that it is physically always present—you cannot enable or disable it. And most importantly, a capacitor has tolerance. A capacitor of 1 nF could have a tolerance of +/− 10%, which means its value could differ from 1 nF by up to 10%. This change could occur due to temperature or simply with age. So, with time, the filtering effect could also change. The controller that receives the final conditioned signal could end up seeing more noise or less noise as the capacitor value drifts.

2.4 The Need for Digital Signal Processing

In some ways, continuous signal processing is quite convenient—cheap and low footprint (size). But as stated in the previous lecture, one big disadvantage is that when performing continuous signal processing, you are the mercy of component values. In some cases, like a simple capacitor acting as a low pass smoothener, drift will not be too much of a problem. For the capacitor to be completely useless, the value of the capacitor would either have to drift by a very large value like 50% or have to be completely damaged.

However, in some other cases, filters can be a bit more complex. For example, let us suppose we are trying to bypass signals of a particular frequency in which case we need a notch filter. An example of such a continuous time notch filter is shown in Fig. 2.2.

Fig. 2.2 A notch filter using passive components

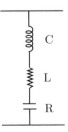

This filter will have a low impedance for a particular frequency (called resonant frequency), and the values of the inductor and the capacitor are usually chosen to achieve this frequency as given by (2.3) [4]

$$f_0 = \frac{1}{2\pi\sqrt{LC}} \tag{2.3}$$

This R-L-C filter will offer an impedance that varies with the frequency of the signals applied across it. The impedance will be minimal for the resonant frequency of (2.3). Signals having frequencies close to the resonant frequency f_0 will be bypassed by the circuit and will not continue into the downstream circuit. For frequencies lower than or higher than f_0, the impedance of the R-L-C filter will be significantly higher than the minimal impedance offered by this circuit, and therefore such signals will not be bypassed by the circuit, and they will continue to flow in the rest of the circuit.

The problem with this kind of filter is quite obvious. What if the value of the capacitor or inductor were to drift? The frequency f_0 would drift, and if this drift becomes significant, the filtering effect would decrease as the impedance offered by the circuit to signals of frequencies close to f_0 would be much higher, and these would now not be bypassed and will continue to flow through the rest of the circuit. The only way to prevent this from happening is to use components with very low tolerance values. As an example, it is possible to choose capacitors with 0.1% tolerance, which means that its value will change by 0.1% at the maximum from the parametric value with temperature and age. However, this means that the cost of the filter will be higher as low tolerance components are typically more expensive.

Another issue is that we cannot selectively change the resonant frequency of the circuit. Of course, theoretically, you could connect series and parallel branches that have additional inductors and capacitors and using switches alter the equivalent inductance and/or capacitance of the circuit. However, this is rarely done in practice. These kind of filters do have their applications in some electrical networks to trap a frequency that causes machines such as transformers and motors to resonate. However, in control circuits, the use of such notch filters is on the decline due to the advent of digital filtering.

In these cases, for the purpose of controls or real-time computations, digital signal processing wins in a big way as will be described in the next section.

2.5 Digital Signal Processing

Digital signal processing (DSP) can be performed in a number of ways using several diverse approaches. As an example, DSP can be achieved by simple digital circuits like flip flops, counters, adders, subtractors, etc. Alternatively, DSP can be achieved using a processor such as a microprocessor/microcontroller, a Field Programmable Gate Array (FPGA) and many others. In this book, when we talk of DSP, we usually imply the use of a processor. The choice is dependent on the budget and the expectations in terms of speed. Moreover, the term DSP will be used for the technique digital signal processing. However, there are also microcontrollers available based on this technique and are called DSP microcontrollers. These microcontrollers use some of the DSP techniques due to which they are named such. In this book, the emphasis is on the technique of DSP rather than the method of implementation.

We will soon talk about how a filter is implemented. For now, imagine that you were to be able to perform filtering by simply writing a computer program. In this case, the processor would receive as an input the signal that needs to be processed—for example, elimination of a particular frequency component. The processor would, by using computations, provide an output where this frequency component has been reduced to the extent of being eliminated. However, since we now have a processor at our disposal, our flexibility in programming it is much greater. For example, it could remove signals of a particular frequency while allowing other signals to pass through. Therefore, the need to connect parallel and series combinations of passive components in order to expand the filtering action is now not necessary. In these modern times, even inexpensive processors are fairly powerful and can perform computations that are quite complicated. The possibilities are endless depending on how powerful your processor is, and many modern microcontrollers allow fairly complex calculations in real-time that even a decade back would need a high-end desktop computer.

Before we get into the details of DSP in terms of how it is implemented, what would be the fundamental difference between continuous time processing and digital processing? We have seen the case of the filter capacitor and the R-L-C circuit before where the filters were always (continuously) acting on the signals available to them. However, when a processor is involved, once a signal is fed to a processor, the computations performed by the processor on that signal would take some time. Again, depending on how powerful your processor is and how complex your computations are, the minimum amount of time needed to perform a certain set of computations will vary. However, it will take some finite non-zero time. During this time, you cannot feed it other signals and expect them to be processed as well. Theoretically, one could argue—what if we had a multi-core processor and we had an advanced operating system that can assign tasks to different cores and have them run computations simultaneously? We could. But these are very advanced concepts, and we will not deal with them in this book. Let us say we have a simple basic processor that can perform one task at a time.

Therefore, the signals that will be fed to the processor will need to be sampled. That means instead of feeding signals continuously, we pick out signals at intervals of time and feed these signals to the processor for computation. And the rest of the time? We do not care. You might ask—why bother with this sampling? Why not just feed all signals to the processor and let it handle what it can? The problem is if we do so, we then may not know at what time instants the signals are being processed and at what time instants they are ignored because the processor is busy. Next question— does it make a difference exactly when the processor performs computations based on a set of sample inputs? Yes. A lot of filter design as we will see in the later chapters does take into account the sampling interval and if this is not adhered to, filters may either not work as expected or even worse may be unstable.

Therefore, in digital signal processing, the first step that needs to be understood is that sampling is critical. Sampling is essential to allow the processor sufficient time to complete computations. However, sampling also plays a critical role in the design of digital filters. The importance of this will be described in the next chapter when we introduce the theory behind digital signal processing. Therefore, digital signal processing when using a digital platform such as a processor is always discrete time signal processing where sampling plays a role in both the design and the implementation stages of a filter.

2.6 Advantages of Digital Signal Processing

Now that we have started with our definition of digital signal processing, let us use the abbreviation DSP from henceforth. Without going into details of DSP so early in the course, let us look at some obvious advantages of DSP. The first is we now have a processor involved where we can write programs. That should bring a smile into so many of our faces as programming is becoming a skill that is ubiquitous with most engineers. If we can harness the capabilities of computers and processors, we have made a huge jump in what we can achieve. This is for the simple reason that software skills are far easier to acquire and hone particularly in this age of online learning as compared to electronics design skills. Therefore, for the sake of simplicity and ease of implementation, DSP scores in a big way.

The second is reliability. When a processor works, it works and the results are repeatable and reproducible. Processors do not suffer from problems like drift. A computation will always produce the same result. There might be some changes with respect to some of the decimal fraction values depending on how many bits our processor has—16, 32, 64 bit, etc. But the values will not simply drift as the processor ages. For a given processor, computations are guaranteed to produce the same results over and over again as long as the processor does not get physically damaged.

Of course, a processor could also fail. After all, everything can and at some point of time will fail. We see that with our computers, our mobile phones—just about any electronic appliance. Someday, they may not boot, the sound may disappear and the screen may start flickering. All machines have their points of failure. And a processor could also have the same. However, failures do not happen all the time and when they do it is obvious. If a computer fails, you would know. On the other hand, a filter based on passive components could be rendered useless due to drift in the component parameters and you may not realize it. The failure of a processor can usually be detected.

Therefore, we have now a more dependable signal processor. However, the cost of our system has increased. And this usually means, if we have more complex processing needs, the additional cost of a processor can be justified. However, in modern systems, there are many applications of advanced signal processing that are welcomed if not downright necessary. And, with the falling cost of microcontrollers and microprocessors, the cost difference is usually not prohibitive.

2.7 Process of Converting a Signal to Digital Form

We already talked about the process of sampling where we choose samples of a signal at discrete time intervals. The next question is how do we feed these samples to a processor for computation? For that, we need to first understand how a processor deals with data. A processor talks in the language of ones (1) and zeros (0). The same way we use words with alphabets, for a processor, there are just ones and zeros as shown in Fig. 2.3.

These 1s and 0s are called binary digits (bits). To start, if someone were to ask you a question and expects only Yes or No as an answer, what that means is the data expected from you has two possibilities only. In a similar manner, if you were to feed either 1 or 0 to a processor, you are feeding a bit. A bit is the most basic form of data for a processor. It is a single digit that can be either 0 or 1—two possibilities only. It is important to note that 0 and 1 are merely representative numbers and not 0 and 1 the way we know it from basic arithmetic. In other words, a bit can be said to represent two levels with 0 being one level and 1 being another level. Let us try to illustrate this with an example of trying to measure temperature in different parts of a room and represent it as bits with the intention of feeding this data to a processor.

Let us assume that this room is fairly big and the temperature across the room varied by 10 °C altogether. Close to the heater, the temperature could be as high as 28 °C. At the other end of the room, away from all heating outlets, the temperature could be as low as 18 °C. This is just one example. The difference in temperature could be higher or lower. To be able to express the worst case, which is the widest possible temperature difference in cases of extreme weather, let us add another 50% to the temperature difference that we have currently assumed. So, our goal should be to measure and feed this temperature to a processor.

Let us start by trying to use a single bit—1 or 0 to represent this temperature difference. The bit 1 could represent the maximum temperature of 30.5 °C, while bit 0 could represent the minimum temperature of 15.5 °C. Anything above 30.5 °C will all be bit 1, and anything less than 15.5 °C will all be bit 0. Between 15.5 and

Fig. 2.3 Binary digits

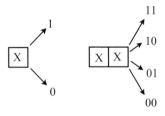

30.5 °C, it could be either bit 0 or bit 1. To differentiate, let us take the mid-point of these two temperatures, 23 °C. Anything lesser than 23 °C is bit 0, and anything greater than 23 °C will be bit 1.

As you can quite obviously see, with just 1 bit, your ability to represent the data will be quite poor and errors are large. We could always add another bit as shown in Fig. 2.3. Each bit has 2 levels. So, there are 4 possibilities—00, 01, 10 and 11. In general, for a binary number with N digits, the number of levels that can be represented is given by the relation:

$$\text{Number of levels} = 2^N \tag{2.4}$$

With 4 levels, we can now divide up the temperature range—15.5, 20.5, 25.5 and 30.5. If you notice, the gap between the levels is total difference between extreme temperatures divided by $N - 1$. Our data representation improves a bit. As before, anything less than 15.5 °C is 00 and above 30.5 °C is 11. Between 15.5 and 20.5 °C is either 00 or 01, between 20.5 and 25.5 °C is either 01 or 10 and between 25.5 and 30.5 °C is either 10 or 11. Again, we use the mid-point method to separate data into levels when it falls between levels.

The levels can be arranged in series by calculating the decimal equivalent of their binary representation. Any general N digit binary number can be represented as shown in Fig. 2.4. The decimal equivalent for such a binary number can be expressed as the following equation.

$$\text{Sequence of level} = b_0 + b_1 \times 2 + \ldots + b_{N-2} \times 2^{N-2} + b_{N-1} \times 2^{N-1} \tag{2.5}$$

where b_x can be 0 or 1. So, 00 is 0 (0th level), 01 is 1 (1st level), 10 is 2 (2nd level) and 11 is 3 (3rd level).

With increase in the number of bits and the subsequent increase in the number of levels, the data representation gets better. As an example, if the number of bits were increased to 12 as shown in Fig. 2.5, the number of levels jumps to 4095. The error

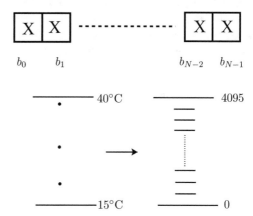

Fig. 2.4 Converting a binary number to a decimal equivalent

Fig. 2.5 Quantization error with 12 bits

between an actual data point and the representative level to which it is assigned to becomes lower. But there is error. And that is important to know, there will always be some error. This error is called quantization error [5].

In this section, we have introduced the concept of quantization by which a real-world signal is represented in digital form. Sampling and quantization are the first steps in moving from the continuous domain to the digital domain. Both sampling and quantization are tasks that are performed by off-the-shelf integrated chips available from several manufacturers. This will be described in the next section.

2.8   Practical Analog to Digital Conversion

Now that we have talked about the concept of converting a continuous analog signal to digital form with the example of temperature measurement, let us look at how it is done practically. You will find the mention of an ADC used frequently, which stands for analog to digital converter. An ADC combines all these functions together—sampling, quantization and final transmission in digital form.

The first—sampling. For this, the ADC has a sample and hold circuit as shown in a simplified manner in Fig. 2.6 [6]. Simply put, there is a switching device that when turned on connects the input to a capacitor. The switching device will turn on at a command. This command is usually provided at constant time intervals equal to the sampling time period. As stated before, the sampling time interval does not have to be constant; however, for any sensible digital system design, a constant sampling interval is necessary. When the switching device turns on, it will remain switched on for a time interval called the hold time. This is to ensure that the capacitor attains a voltage equal to the input voltage. This hold time is essential because the voltage across a capacitor will not change instantaneously, and the capacitor needs to charge or discharge to attain the voltage equal to that applied across it.

After this hold time, the switch will turn off, and the voltage of the capacitor remains constant and is an approximation of the signal that has been sampled. Now, the process of quantization follows. Here, we bring in another definition—the resolution of the ADC. The resolution of the ADC is usually specified as a number of bits—for example, 10, 12 bit, etc. The number of bits of the ADC denotes the number of bits in the final digital output or representation of the signal sample.

Fig. 2.6 Sample and hold circuit

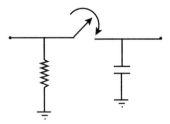

Typically, the most commonly used is the 12 bit ADC. Lower resolution ADCs have 8 bits or 10 bits. 16 bit ADCs are also fairly common. ADCs with 24 bit resolution and higher are usually special applications such as military or aerospace where accuracy is important and these are usually more expensive. The number of bits translates to the number of quantization levels. 12 bits implies $2^{12} = 4096$ as per our previous calculation of determining the number of distinct levels for a certain binary number. This in turn translates to 4095 levels (minus 1). With such a number of quantization levels, the accuracy of representing an analog signal in digital form is usually acceptable for most applications in automation and industry.

Though it is not possible to examine how a sampled signal is assigned to a particular level, it would be necessary to fill this gap in this section. In order to perform this assignment, the ADC has a number of comparators that compare the signal value with references that correspond to each level. These voltage references are generated from the power supply provided to the ADC chip and with respect to the voltage range of the ADC. As an example, most ADCs that are powered by a supply of 3.3 V will specify their input range to be 0 to 3 V. This implies that any signal connected to the input must be within this range. By specifying this range, the ADC internally divides the voltage range into a number of levels as required by the final resolution. Therefore, a 12 bit ADC will divide 0 to 3 V into 4095 levels. In order to assign a sampled input signal to one of these levels, there will be a vast number of comparators. And therefore, increasing the resolution will increase the cost of the ADC as the complexity of the hardware increases with a greater number of comparators.

The final result will be stored in a result register and made available to a processor. As can be seen, the tasks involved in the conversion process are well-timed. Until the conversion is complete, the result is not available and what might be present in the result register should not be used. Similarly, the conversion should start when the previous result has been used or we run the risk of overwriting results. In the next section, we will describe how the processor and the ADC are interfaced.

2.9 How Would a Processor Handle the ADC?

In the previous section, we presented the ADC and we saw that an ADC samples a signal and converts it to a digital number. The next question is how is this process regulated? This process is usually regulated by the processor, and in this section, we will look into this coordination between the ADC and the processor.

Figure 2.7 shows how the processor and the ADC communicate [7]. To begin with, the ADC needs a command to begin sampling. The processor decides through control code, how often sampling should occur and provides the signal to start sampling, which the ADC uses to turn on the sampling switch. This is usually called the start of the conversion (SOC) process. When the ADC completes the quantization and then converts the signal to the 12 bit binary number, this number gets stored in a result register. To indicate that the ADC has a result available, it

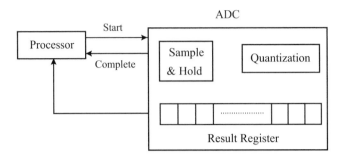

Fig. 2.7 Block diagram showing the communication between the processor and the ADC

will provide a signal to the processor, and this signal is usually called the end of conversion (EOC). In hardware, this is usually done by a pin of the ADC going high (or low), and if this pin is connected to a pin of the processor, the processor reads the change of state of the pin. The processor knows data is available and reads the converted signal stored in the result register.

This entire process gets repeated the next time sampling needs to be done. Typically, to ensure that this processor is done at constant time intervals, a timer is used. At completion of a time interval, the timer usually produces an interrupt. The interrupt is a signal to the processor from a peripheral (in this case, the timer) that an event has occurred, and the processor must stop its usual processes to service the event. Such a timer interrupt can be used to ensure that the processor issues a command to start the conversion. Similarly, when the ADC completes the conversion, the EOC signal issued can be made to trigger another interrupt to indicate that another peripheral (the ADC) has produced an event, which in this case is that the digital result of the sampled signal is ready. The processor can read the result register of the ADC and use the value for computations. This is of course a simple example, and in some cases, there are more advanced techniques employed.

2.10 Conclusions

This was a fairly light chapter to begin with. The objective was to talk about digital signal processing from a practical perspective rather than jumping straight into theory. As power engineers, hardware is something we are always comfortable with and so, I thought, best to get started with hardware.

The main concepts to remember from this chapter are that sampling is absolutely critical to the concept of digital signal processing (DSP). In the later chapters, sampling will be used to represent a system mathematically, and it will be evident that the sampling time interval will play a crucial role in the behaviour of systems. DSP does not need a processor and can be achieved using digital circuits; however, this course will assume a processor as in the later chapters, and we will look at programs using Python where a filter will be implemented.

Though we started with hardware, the rest of the book will not contain any hardware implementation. The focus of the book is on design, analysis and the conceptual implementation of filters. When I mean conceptual implementation, you will find in the later chapters that the code used to implement the filter can be translated to hardware on a microcontroller with minimal changes.

References

1. Oppenheim, A. V., & Schafer, R. W. (2001). *Discrete-Time Signal Processing*. Pearson.
2. Stranneby, D., & Walker, W. (2004). *Digital Signal Processing and Applications* (2nd ed.). Elsevier.
3. Ramsden, E. (2006). *Hall-Effect Sensors: Theory and Applications* (2nd ed.). Elsevier.
4. Rao, B. V., Rajeswari, K. R., Pantulu, P. C., & Rama, K. B. (2012). *Electronic Circuit Analysis*. Pearson Education India.
5. Widrow, B. (1956). A study of rough amplitude quantization by means of nyquist sampling theory. *IRE Transactions on Circuit Theory, 3*(4), 266–276.
6. Candes, E. J., & Wakin, M. B. (2008). An introduction to compressive sampling. *IEEE Signal Processing Magazine, 25*(2), 21–30.
7. Walden, R. H. (1999). Analog-to-digital converter survey and analysis. *IEEE Journal on Selected Areas in Communications, 17*(4), 539–550.

Chapter 3
Getting Started with Signal Processing

3.1 Introduction

In the previous chapter, we introduced DSP using a few non-engineering and some well-known electrical engineering examples. We had also compared DSP with analog signal processing with advantages and disadvantages of both. With an overview of DSP, in this chapter, let us dive into signal processing theory.

In this chapter, we will start with circuits made of passive components such as inductors and capacitors and look at their properties as filters. We will analyze basic filters such as the LC filter using network laws to be able to lay the foundation of the frequency dependent behaviour of circuits. We will then present the Laplace Transform and examine its properties in transforming common electrical signals as well as common circuit laws. We will examine the impact of Laplace Transform and the significance of the resultant function. We will then examine how to convert a system from the continuous domain to the digital domain.

This chapter will present theory with basic examples but will not get started with the actual filtering or the performance of filters. The focus of this chapter is to understand how starting from a time domain representation of the system, we can arrive at a frequency domain representation of a system. This frequency domain representation of a system is the fundamental basis of signal processing as it allows us to analyze the behaviour of the system for different frequencies.

3.2 Reviewing Capacitors and Inductors

We saw in Chap. 2, a few basic filter examples of how we used capacitors and inductors as filters. Let us continue with analog filters such as these but try to take a step back and look at the process mathematically instead. Take for instance the

S. V. Iyer, *Digital Filter Design using Python for Power Engineering Applications*, https://doi.org/10.1007/978-3-030-61860-5_3

capacitor filter. The primary equations of a capacitor are [2]:

$$i = C\frac{dv}{dt} \tag{3.1}$$

$$v = \frac{1}{C}\int i\,dt \tag{3.2}$$

Let us try to map the physical properties of the capacitor to these equations. The current through the capacitor branch being a derivative with respect to time of the voltage across the capacitor implies that the current will increase as the voltage changes rapidly. As an extreme case if the voltage is constant—a dc voltage—the current will be zero, as the voltage does not change at all, and so the rate of change is zero. Physically, if you apply a dc voltage across a capacitor, it will charge up to a voltage equal to the applied voltage after which the current will be zero. Take a very high frequency noise signal as the applied voltage as the other extreme case. Now, the rate of change is very high simply because it has a high frequency to begin with. The current through the capacitor will be high and will be a high frequency component as well. The capacitor will therefore block a dc current and allow a high frequency current to pass through.

What about the integral relationship in (3.2)? The exact opposite. The integral is the equivalent of a sum over time. Therefore, you are accumulating with respect to time, which automatically has a smoothening effect. Over time, the small events lose importance and the major events are what matter. As an example, you will not remember each time you paid your mobile bill, but you will remember the time you bought your mobile. Because the former are minor events and the latter is a major event. Therefore, the voltage across a capacitor will be much smoother than the current flowing through the capacitor branch.

Let us look at an inductor. The equations for the inductor are quite the opposite of a capacitor [2]:

$$v = L\frac{di}{dt} \tag{3.3}$$

$$i = \frac{1}{L}\int v\,dt \tag{3.4}$$

From the differential relation of (3.3), the voltage across the inductor will be directly proportional to the rate of change of current through it. And like the previous case, the extremes are interesting. If the current is constant—a dc—the voltage will be zero, because the current never changes and the rate of change is therefore zero. If the current is a high frequency noise, the voltage will be high simply because high frequency means high rate of change.

It is not immediately obvious what the voltage in (3.3) physically signifies. The voltage in (3.3) is actually the induced emf produced by the inductor. This induced emf by Lenz's law always opposes the cause that produces it. Therefore, the voltage across the inductor in (3.3) is the opposition that the inductor offers to the external circuit. So, if the voltage is zero, the inductor offers no opposition (zero impedance), while if the voltage is high, the inductor ferociously opposes the external circuit that is trying to push a current through it. So, an inductor blocks high frequency signals and lets low frequency signals pass through more easily.

Now, let us examine the integral relationship of (3.4). As before, the integral has a smoothening effect. So, if the voltage is high frequency, the current will reduce the effect of the high frequency components. If the voltage is low frequency, the current could build up to fairly large values. The extreme case is if the voltage is a dc, in which case the current will continuously increase. In practice, the inductor will saturate and will probably burn after a limit.

You will find from the above discussion that capacitors and inductors behave almost like opposites in any circuit. Capacitors block dc currents and pass high frequency currents, while inductors pass dc currents and block high frequency currents. These properties cause capacitors and inductors to go hand in hand in almost every circuit as filters, as will be shown soon.

3.3 Filters as Combinations of Inductors and Capacitors

In the previous section, we concluded about how capacitors and inductors can be used as filters. In this section, we will examine filters constructed from inductors and capacitors. As a starting example, if we had a signal and we wish to smoothen it out—which means remove/reduce high frequency components—a typical filter would be an LC filter as shown in Fig. 3.1. In such a filter, an input v_{in} is a signal that could potentially contain high frequency components that we wish to remove.

Let us break this circuit up and examine the behaviour of the inductor and capacitor separately. The behaviour will be obvious once we write the equations that describe the circuit. The equations can be written as [2]:

$$i = \frac{1}{L} \int (v_{in} - v_o) \, dt \qquad (3.5)$$

Fig. 3.1 LC filter

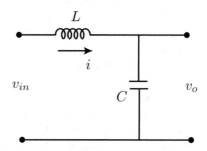

$$v_o = \frac{1}{C} \int i \, dt \tag{3.6}$$

From (3.5), the inherent property of the inductor is evident where due to the integral operation, the current will be a smoother signal as compared to the input voltage, and high frequency components present in the input will be attenuated in the current. This follows from our conclusion in the previous section that the inductor blocks high frequency components and lets low frequency components pass through. This fits in well with our objective of designing a filter that removes high frequency components present in the input.

From (3.6), the inherent property of the capacitor is similarly in action. Mathematically, the integral operation will result in an output signal v_o that has a much lower high frequency component than in the current i. Given the fact that the current has already been filtered by the inductor and in turn has a lower high frequency component than the input, the output v_o will have a significantly lower high frequency component than the input v_{in}. You could see that a double filtering is seen to take place.

Here, there might seem to be a bit of a puzzle. What about the property of the capacitor that allows high frequency components of current to pass through and blocks low frequency components? With the inductor, the blocking nature stopped high frequency components, which was in line with our filtering objective. However, with the capacitor, there seems to be a conflict. In reality, this is not a conflict but is also assisting in our filtering objective.

The inductor forms the series connection between the input and the output. Therefore, blocking high frequency components is an obvious benefit. The capacitor, on the other hand, is connected in parallel with the output. It is to be noted that in Fig. 3.1, the output terminals are shown as open terminals. In reality, another circuit will be connected across those terminals. The capacitor allows high frequency components in inductor current to pass through it, thereby bypassing any high frequency noise that may remain in the current i through the inductor. Therefore, the circuit connected in the output is shielded from these high frequency components. This is also in line with our filtering objective—we want to stop high frequency currents from reaching the output.

Next question—if the current through the capacitor is primarily high frequency components, how is the output voltage devoid of these high frequency components? To figure this out, look at (3.6). Due to the integral operation, even if the current through the capacitor has high frequency components, the output voltage will see them attenuated. And this is the reason, why in many circuits, you will find capacitors connected across nodes where the voltage needs to be smoothened. The capacitor bypasses high frequency components but still produces a smooth voltage across it.

And this is what was the meaning of saying that inductors and capacitors go hand in hand while designing filters. The inductor plays the first role of blocking high frequency components. Those high frequency components that remain are then bypassed by the capacitor, which acts as some kind of decoy while also producing a fairly smooth voltage across its terminals for an output to be connected. In the later sections, we will examine the operation of the LC filter in greater detail when we design and simulate it.

To conclude our analysis, the equation of the output with respect to the input is

$$v_o = \frac{1}{L\,C} \int \int (v_{in} - v_o)\, dt \qquad (3.7)$$

It is possible to solve this equation mathematically and of course numerically. However, in practical circuits, it is quite common to design higher order filters. As an example, it is quite common to stack filters up to form a ladder. In this case, the filtered output of the first is fed to the second, and so on. Therefore, the integrals will keep stacking up as well. How do we solve this?

We could just go numerical in every case. I mean why not? We have computers. Just write programs to do the integral or use any of the commercial packages, feed the equations in and it will give you equations or waveforms and you can check out it in a simulation. Unfortunately, this is not a method that many engineers are comfortable with as it is heuristic. The problem with this method is, you cannot perform any analysis on it. The good old solving equations are not possible.

So, the question is how do we analyze higher order filters? One answer is using transformations and we will examine that in the next section.

3.4   The Concept of Transformations

In engineering and science, we always have equations. We love equations, because they give us the opportunity to solve them. Equations to scientists and engineers are like food to a foodie. Equations help to quantify physical phenomena and express processes in an abstract manner. Without equations, it is impossible for science to grow. Usually, we see mathematics and equations as a nuisance that we need to live with. What we are interested in is the final product—something that provides some kind of service. This is unfortunately because mathematics is presented on its own rather than as a tool to achieve a result. In the next few sections, theory will be presented with a constant emphasis to answer the question—how is this of any use?

With equations, there are many different types—polynomial equations, ordinary differential equations, partial differential equations and so on [3]. The list is endless. Every type of equation has interesting properties and makes them easy to solve or extract some specific trait of the system that they represent. It is important to note here that almost no type of equation is useless. Some poor mathematician has probably gone crazy or killed himself or herself studying that type of equation, and there exists literature that we can refer to make inferences when we encounter that type of equation.

When you have an equation in one form, you could solve it or analyze it and get some information and make some conclusions. But does it end there? If it was possible to transform this equation into an entirely different form, this process could be repeated—solve, analyze and conclude. Since the equation is of a different type, the conclusions might be totally different. Theoretically, there is no end to the number of times you could transform equations and repeat this process. Each transformation may yield drastically different conclusions or maybe might just offer a minor additional insight. It is hard to predict what could be the outcome. But one thing is for sure—knowledge accumulates.

And this is why we engineers and scientists like transformations as well. Because they help us to continuously extract information from systems. During the course of engineering, we are presented with several of these—Laplace Transform, Fourier Transform, Clarke's Transformation and Park's Transformation are just a few examples. Unfortunately, transformations are not presented very well or they are presented in their raw mathematical form that simply scares most of us. Unless we go back and forth to examine how a transformation would help in attaining a desired objective, the theory behind the transformation is eventually just forgotten.

I will be presenting soon the Laplace Transform. When I was a student, the Laplace Transform was presented as an equation—we struggled to solve it, we were given a list of properties, which we also struggled to solve. In the end, we did what we could to pass the exam. And we forgot what we learned and life went on. There is a saying that it is better to have learned and forgotten than to have never learned at all. I would say that is most unfortunate. Rather than learning with no hope of remembering, why not learn so as to carry it with you forever?

So, I will go a bit slow with some of these concepts. Also, what I cover may not be strictly mathematically sound. But, the idea is to present signal processing in a way that is easy to understand and remember and most of all to enjoy this wonderful subject.

3.5   Laplace Transform

Proposed in around 1785 by Pierre-Simon Laplace, the Laplace Transform was one of the most transformative operations in the history of science and engineering. Laplace made massive contributions to mathematics and physics in the late 1700s and early 1800s. Those were drastically changing times in France and also, during the Napoleonic era, mathematics was hugely popular and encouraged since Napoleon himself was a lover of mathematics. In all fairness, Laplace was not the very first mathematician to have thought about such a transformation. Work in this area had started in 1744, by Euler and later by Lagrange in trying to solve differential equations by some kind of transformation instead of the usual process of integration.

The Laplace Transform is the following integral [4]:

$$\mathcal{L}\{f(t)\} = F(s) = \int_0^\infty f(t)\, e^{-st}\, dt \tag{3.8}$$

Let us describe this equation in a bit of detail. The symbol $\mathcal{L}$ usually denotes the Laplace Transformation being applied to a function of time $f(t)$. In many cases, it is quite normal to denote the time varying function by small case letters f and the transformed function with the upper case letters F. However, this is just a convention followed in a large number of sources and need not be the case. The Laplace Transform performs a definite integral from 0 to infinity, ∞, over time t. There are other forms of Laplace Transform where the integration is performed from $-\infty$ to ∞ [5]. However, we will consider in this book, the definite integral from 0 to ∞ over time t. The integration is performed on product of the time varying function $f(t)$ and the exponential e^{-st}. The variable s is the frequency, and a later section in this chapter is devoted to describing this variable. The final result of this transformation is a function in the variable s.

In very simplistic terms, the Laplace Transform converts a function in time t to a function in frequency s. For just a bit, take it for granted that s is the frequency until we reach the detailed dig into this mystery. To begin with, you might think, what is the difference? We are performing another integration, so how is this any different from solving a differential equation by integration?

This transformation can be applied to any function of time $f(t)$, and this function could be a differential equation, an integral equation or just about any type of equation. And this is important—any function of time. We are not limited to just solving differential equations. Before we examine what is s and why that matters in signal processing, let us look at what Laplace Transform produces when applied to many functions in time. It is fantastic when you learn that for a large number of functions $f(t)$, the result of the Laplace Transform is an incredibly simple polynomial in s.

To illustrate this, let us look at some examples. For us electrical engineers, there are a few functions that are most important. When we consider signals, the most important signals are the sine and cosine. Let us derive the Laplace Transform for these functions.

$$\text{Suppose } f(t) = \sin \omega t \tag{3.9}$$

Using the Laplace integral:

$$F(s) = \int_0^\infty \sin \omega t \, e^{-st} \, dt \tag{3.10}$$

Does not seem very helpful. It will be easier to perform the integral if we express the sinusoid as exponentials. This is a well known trigonometric relation:

$$\sin \omega t = \frac{e^{j\omega t} - e^{-j\omega t}}{2j} \tag{3.11}$$

It is very easy to figure this out once you use Euler's equation:

$$e^{j\omega t} = \cos \omega t + j \sin \omega t \tag{3.12}$$

With this, the Laplace Transform integral becomes

$$F(s) = \int_0^\infty \frac{e^{j\omega t} - e^{-j\omega t}}{2j} e^{-st} \, dt \tag{3.13}$$

which can be simplified as

$$F(s) = \int_0^\infty \frac{e^{-(s-j\omega)t} - e^{-(s+j\omega)t}}{2j} \, dt \tag{3.14}$$

The integral of an exponential is

$$\int_{t_1}^{t_2} e^{\alpha t} \, dt = \frac{e^{\alpha t_2}}{\alpha} - \frac{e^{\alpha t_1}}{\alpha} \tag{3.15}$$

So, the integral of the sinusoid is

$$F(s) = \frac{1}{2j} \left(\frac{1}{s - j\omega} - \frac{1}{s + j\omega} \right) = \frac{\omega}{s^2 + \omega^2} \tag{3.16}$$

Equation (3.16) shows how a function like the sinusoid has been converted to a second order polynomial using Laplace Transform.

Exactly the similar method can be undertaken for the cosine.

$$\text{Suppose } f(t) = \cos \omega t \tag{3.17}$$

Again, using the Laplace integral,

$$F(s) = \int_0^\infty \cos \omega t \, e^{-st} \, dt \tag{3.18}$$

As before, using Euler's equation,

$$\cos \omega t = \frac{e^{j\omega t} + e^{-j\omega t}}{2} \tag{3.19}$$

With this, the Laplace Transform integral becomes

$$F(s) = \int_0^\infty \frac{e^{j\omega t} + e^{-j\omega t}}{j} e^{-st} \, dt \tag{3.20}$$

which can be simplified as

$$F(s) = \int_0^\infty \frac{e^{-(s-j\omega)t} + e^{-(s+j\omega)t}}{2} \, dt \qquad (3.21)$$

Therefore, the integral of the cosine is

$$F(s) = \frac{1}{2} \left(\frac{1}{s - j\omega} + \frac{1}{s + j\omega} \right) = \frac{s}{s^2 + \omega^2} \qquad (3.22)$$

Besides the ability of the Laplace Transform to convert many signals in time to polynomials in s, another benefit is the ability of the Laplace Transform to transform electrical circuits. In the next section, we will describe how Laplace Transform is applied to circuits.

3.6 Revisiting Inductors and Capacitors with Laplace Transform

Laplace Transform also helps to transform the way we perceive circuits [6]. The base building blocks of most power electronic circuits are inductors and capacitors. Let us begin by examining the equations of the inductor and capacitor:

$$v(t) = \frac{1}{C} \int i(t) \, dt \qquad (3.23)$$

$$i(t) = C \frac{dv(t)}{dt} \qquad (3.24)$$

$$i(t) = \frac{1}{L} \int v(t) \, dt \qquad (3.25)$$

$$v(t) = L \frac{di(t)}{dt} \qquad (3.26)$$

If we would like to perform Laplace Transform on these equations, we end up with two calculus operators. For example,

$$V(s) = \int_0^\infty \frac{1}{C} \int i(t) \, e^{-st} \, dt \, dt \qquad (3.27)$$

or

$$I(s) = \int_0^\infty C \frac{dv(t)}{dt} e^{-st} \, dt \qquad (3.28)$$

So, let us first examine the generalized case of the double integral:

$$f(t) = \int g(t) \, dt \tag{3.29}$$

So,

$$F(s) = \int_0^\infty \int g(t) \, e^{-st} \, dt \, dt \tag{3.30}$$

To begin with, we need to differentiate between the variables in the two integrals. Giving them both the same time t is confusing:

$$F(s) = \int_{t=0}^\infty \int_{\tau=0}^t g(\tau) \, d\tau \, e^{-st} \, dt \tag{3.31}$$

What exactly are we doing here? The inner integral in the variable τ produces an equation in t, and the outer integral takes this equation in t and integrates that from 0 to ∞. Eventually, we are integrating from 0 to ∞. This is the most important to understand. So, as long as we do so, we can shuffle the integrations a bit. Suppose, we integrate the inner one from t to ∞ and the outer one stays from 0 to ∞. It would be the same thing.

$$F(s) = \int_{t=0}^\infty \int_{\tau=t}^\infty g(\tau) \, d\tau \, e^{-st} \, dt \tag{3.32}$$

Now, there is another cool trick we can use. We simply rename our integral variables because in the end, all that matters is that we perform the integral over the limit.

$$F(s) = \int_{\tau=0}^\infty \int_{t=\tau}^\infty g(\tau) \, d\tau \, e^{-st} \, dt \tag{3.33}$$

This might seem like a useless operation, but this makes it possible to split up our integration order very conveniently. We can now separate the integrals:

$$F(s) = \int_{\tau=0}^\infty \left(\int_{t=\tau}^\infty e^{-st} \, dt \right) g(\tau) \, d\tau \tag{3.34}$$

This is just because the exponent is such an easy function to integrate. If you see an exponential, try to integrate it and be done with it. This gives you:

$$F(s) = \int_{\tau=0}^\infty \left(\frac{e^{-s\tau}}{s} \right) g(\tau) \, d\tau \tag{3.35}$$

Here, s is just a constant and can be taken out of the integral.

$$F(s) = \frac{1}{s} \int_{\tau=0}^{\infty} e^{-s\tau} g(\tau) \, d\tau \tag{3.36}$$

And the inner integral is nothing but the Laplace integral! Remember, the names of the variables do not matter, you can replace t by τ or by α. So,

$$F(s) = \frac{G(s)}{s} \tag{3.37}$$

Now, let us figure out what to do when

$$f(t) = \frac{dg(t)}{dt} = g'(t) \tag{3.38}$$

As before,

$$F(s) = \int_{0}^{\infty} g'(t) e^{-st} \, dt \tag{3.39}$$

When we have something like this, we can integrate them by parts. As an example,

$$\int_{a}^{b} u(x)v'(x) \, dx = [\, u(x)v(x) \,]_{a}^{b} - \int_{a}^{b} u'(x)v(x) \, dx \tag{3.40}$$

This is a fairly well-known technique in calculus. Using this gives us

$$F(s) = [\, g(t)e^{-st} \,]_{0}^{\infty} - \int_{0}^{\infty} g(t) \frac{d(e^{-st})}{dt} \, dt \tag{3.41}$$

$$F(s) = \frac{g(\infty)}{e^{\infty}} - g(0) - \int_{0}^{\infty} g(t)(-se^{-st}) \, dt \tag{3.42}$$

Here, we need to make an assumption. If $g(\infty)$ is bounded, which means that it is not an unstable system that is continuously increasing with time but instead eventually settles to a reasonable value, the first term will become zero due to the e^{∞} in the denominator. As for the last term, s is again just a constant that can be taken out of the integral. So, the simplified equation is

$$F(s) = -g(0) + s \int_{0}^{\infty} g(t) e^{-st} \, dt \tag{3.43}$$

And just like before, the integral is the Laplace integral. So,

$$F(s) = sG(s) - g(0) \tag{3.44}$$

We now have generalized equations for finding the Laplace Transform for functions that are themselves integrals and derivatives of other functions. We can simply use this for inductors and capacitors. For a capacitor,

$$v(t) = \frac{1}{C} \int i(t)\, dt \tag{3.45}$$

$$V(s) = \frac{I(s)}{sC} \tag{3.46}$$

$$i(t) = C\frac{dv(t)}{dt} \tag{3.47}$$

$$I(s) = sCV(s) - Cv(0) \tag{3.48}$$

And for the inductor,

$$i(t) = \frac{1}{L} \int v(t)\, dt \tag{3.49}$$

$$I(s) = \frac{V(s)}{sL} \tag{3.50}$$

$$v(t) = L\frac{di(t)}{dt} \tag{3.51}$$

$$V(s) = sLI(s) - Li(0) \tag{3.52}$$

Laplace Transform has converted integral and differential equations for the inductor and capacitor to simple polynomials in s. We will examine in the later section, how the Laplace Transform can be used to derive the input–output relationship for filters comprised of inductors and capacitors using the LC filter as an example.

3.7 What About the Original Variables?

In the previous section, we looked at how Laplace Transform can be used to transform functions in time to simple polynomial equations in s. Many questions should have arisen in any reader's mind? What are we doing? We are holding up the Laplace Transform as a magic wand that will help us in dealing with electric circuits and signals. The Laplace Transform itself is an integration, and we saw a few examples in the previous sections where we applied it to signals like the sine and cosine or to the inductor and capacitor equations.

The first question is, is there any real benefit here? We are after all performing an integral and now we are performing an integral even when no differential equations are present. Seems like we have made things worse—before an integration was necessary when there was a differential equation. The second questions is, by transforming a function of time t to frequency s, are we losing or gaining information? Is this process reversible and is there a way back—can we transform back from s to t?

Here, we need to go back to the previous discussion on the need for transformations. Transformations merely convert equations of a particular type to another type. The new type of equations can be used to perform a whole new analysis as opposed to the original set of equations. No equation is useless. They are just different. A function in time can be used to examine how the function behaves at a particular instant of time. A function in frequency can be used to examine how the function behaves at a particular frequency.

Functions in the frequency s can be transformed back to functions in time t using the inverse Laplace Transform, which is [7]:

$$f(t) = \mathcal{L}^{-1}\{F(s)\}(t) = \frac{1}{2\pi j} \lim_{T\to\infty} \int_{\gamma-jT}^{\gamma+jT} e^{sT} F(s)ds \qquad (3.53)$$

Seems like a very complex equation, but not to worry. For us engineers, we do not have to solve this as there are usually standard lookup tables to convert equations in s to expressions in t.

Moreover, quite often we rarely will ever need to convert functions back to the time domain. This is something that we will see in the later sections when we examine filter design and frequency domain analysis. After performing Laplace Transform and designing our filters, there exists very well-established theory to be able to deduce the performance of our filters without needing to convert them back to the time domain. Additionally, we will examine methods of implementing filters digitally from the functions in frequency s. We will also verify the performance of these filters in the time domain through the process of simulation. Therefore, never in this book, we will ever be performing inverse Laplace Transform.

For a vast number of applications, the Laplace Transform can be treated as a one-way street. To convert an equation in time to an equation in frequency. In the next section, we will uncover another question that has been lingering for a while—how is this variable s the frequency?

3.8 The Mystery Behind s

Now that we have presented the Laplace Transform as a mathematical operation, let us look at the physical significance of this operation. Most importantly, we will answer two questions—what is this variable s that the Laplace Transform produces and how is s the frequency? Let us look at the factor that brings in s:

$$e^{-st} \qquad (3.54)$$

This term appears in all the Laplace Transform integrals above. Let us dig into this term.

The variable t is of course the time in seconds. Before we describe how s is the frequency, let us state that the unit of s is per second or /second. Therefore, the product st is dimensionless. We engineers usually expect frequency f to be in hertz. Hertz essentially is cycles per second, so 50 Hz means 50 cycles per second. However, in electrical engineering, we also use another definition of frequency called the angular frequency denoted by ω. This is $\omega = 2\pi f$ and its dimension is radians per second. As radian does not have a unit, angular frequency has a dimension of /s or just per second. So, this in a way explains how s, if it was frequency, can have a dimension of per second. We still have not come to how s is the frequency.

Before we discuss frequency or angular frequency, we need to look at what is a radian and how the dimension of angular frequency is radian per second. We are used to expressing angles in degrees (°)—30°, 90°, 180° etc. This expression of angle in degrees would be considered the layman's expression of an angle. However, in mathematics, angles are expressed in radians instead of degrees. When we use the sine, cosine or any other trigonometric function, the angle passed as the argument is expected to be radians. To further understand how this dimension radian came into being, we need to take a short historical tour.

To express an angle in radians, we need the mathematical constant π. π has a long history and has fascinated a lot of people including inspiring a Hollywood movie. However, the first usage of the constant π was to be able to express the circumference of a circle with respect to its diameter as shown in Fig. 3.2 [8]. It might seem trivial now in the twenty-first century. But, when this was first thought of, it was a revolutionary idea—to be able to relate two very different aspects of a circle. Let us look at a primitive use of this. If you wanted to dig a well, you could decide how wide (the diameter) of the well would be and could calculate the circumference to find out the materials you need to build this well—bricks, mortar etc. Because, the material would need to be purchased with respect to the circumference of the well. But when you dig the well, it is the diameter that you can directly control.

This mathematical constant π has a value equal to $\frac{22}{7}$. It is quite popularly cited as an example of an irrational number. An irrational number is a ratio of numbers that will never result in a remainder of zero when performing the division. So, you

Fig. 3.2 The origin of π

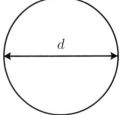

Fig. 3.3 Using π to express angles

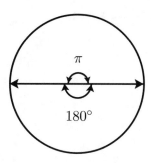

could divide 22 by 7 and you would never ever stop. π is a constant, and so it has no dimension. However, another use for π is to be able to express angles. When used to express angles, the resultant angle is given a nominal dimension of radian. The reason why I say nominal dimension is because radian is not really a dimension. An angle expressed in radians is just a number, but since we are expressing an angle that is a physical property, it would make sense to give the angle a dimension even if it is really a pseudo-dimension.

To understand how π can be used to express angles in radians, check out Fig. 3.3. Since π is used to connect the diameter and the circumference of a circle, another way to look at π is to look at the angle mapped out by the diameter. From Fig. 3.3, the equivalence between π and 180° is fairly obvious. And when expressing angles in radians, it is quite customary to convert an angle in degrees (°) to an angle in radians by multiplying by the factor of $\frac{\pi}{180}$. Now that we have looked at how angles are expressed in radians, let us next look at how sinusoids and cosinusoids are expressed.

The sine and cosine of an angle θ can be expressed as follows:

$$\sin\theta \quad , \quad \cos\theta \tag{3.55}$$

It should be noted that the angle θ is usually in radians and if in degrees, it is converted to radians by multiplying by the factor of $\frac{\pi}{180}$. Though as electrical engineers, the above sin and cos expressions are the ones we use all the time, mathematicians prefer another expression. In mathematics, the most popular way of expressing a sine or cosine is using Euler's equation [9]:

$$e^{j\theta} = \cos\theta + j\sin\theta \tag{3.56}$$

Here, $j = \sqrt{-1}$ is the imaginary number. This relation was proved by the mathematician Euler and thus the name. If you take the exponent of an angle as an imaginary number, the result is a complex number with the real part being the cosine of the angle and the imaginary part being the sine of the angle. This might seem like an unnecessary complication, but the beauty of Euler's equation and expressing a sine and cosine together will be immediately obvious once we describe it pictorially.

Figure 3.4 shows a circle. For simplicity, since we are looking only at angles, the radius of the circle is unity. The circle is in a two dimensional plane with two

Fig. 3.4 Euler's equation

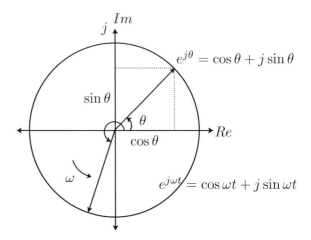

axes called *Re* and *Im*. *Re* that is along the x-axis or the horizontal axis is the real axis, while *Im* that is along the y-axis or the vertical axis is the imaginary axis. This renaming of the x- and y-axes as the real and imaginary axes, respectively, is to be able to express complex numbers since the Euler's equation produces a complex number. In the first quadrant is shown a vector, which makes an angle θ with the positive real axis. It should be noted that the angle that any vector makes is always measured with respect to the positive real axis and measured starting from the real axis and going counter clockwise until we reach the vector. With this definition, the projection of the vector on the real axis is the cosine, while the projection on the imaginary axis is the sine.

So, using Euler's equation, any vector's projection on the real axis and imaginary axis can yield the cosine and sine of the angle of the vector. And these two trigonometric values can be expressed with a single exponent $e^{j\theta}$, where the power of the exponent is an imaginary number. Let us now suppose that this vector is rotating in the counter clockwise direction at an angular frequency ω. At any time t, the angle made by the vector with the positive real axis is ωt. An arbitrary vector is shown in the third quadrant. For this vector, the exponent $e^{j\omega t}$ will result in this relation:

$$e^{j\omega t} = \cos \omega t + j \sin \omega t \tag{3.57}$$

At this point of time, we will stop saying angular frequency and just say frequency, which is what ω is mathematically. Since the resultant angle is in radians, the dimension of frequency ω is therefore quite obviously radians per second. So, if we go back to how we arrive at mathematical frequency from the electrical frequency in hertz:

$$\omega = 2\pi f \tag{3.58}$$

If the electrical frequency f was 50 Hz, which is 50 cycles per second, ω would be 100π radians per second. The physical significance of this conversion is as follows—a single cycle of a 50 Hz sine wave spans 2π radians or $360°$. Therefore, in one second, 50 cycles would span 100π radians. Any electrical frequency in hertz could be converted to radians per second equivalent. However, another result of Euler's equation is that we now have another way to express cosine and sine of an angle, and we will now see how this resembles the variable s that resulted from Laplace Transform.

Let us place the two exponents together—the first is the term in the Laplace Transform and the second is the result of Euler's equation:

$$e^{-st} \quad , \quad e^{-j\omega t} \tag{3.59}$$

Here, $e^{-j\omega t}$ is the complex conjugate of the original Euler's equation, which merely means the imaginary part changes sign.

$$e^{-j\omega t} = \cos \omega t - j \sin \omega t \tag{3.60}$$

From the comparison, it is immediately clear that s is analogous to $j\omega$ and e^{-st} is simply an expression of the cosine and sine of the angle st with s being the frequency. So, Laplace Transform converts a function in time t to a function in frequency s by multiplying the function in time with a sinusoid of frequency s expressed using Euler's equation. Due to the definite integral in the Laplace Transform, the resultant function is devoid of time t but contains the frequency s of the sinusoid.

The next obvious question would be why then are we using a variable s when mathematically we call frequency ω? This is because s is itself a complex number and has two parts:

$$s = \sigma + j\omega \tag{3.61}$$

s therefore is a complex frequency. This at first is hard to comprehend when you come from a hardcore engineering background where frequency is in hertz and there is not much that is imaginary about it. But the reason why s is allowed to be complex is because in signal processing, a frequency domain representation of a system could contain a polynomial in s that could have roots that need not always be real. For that matter, quite often when you solve equations, sometimes you may not be able to arrive at a real solution and the solution might end up being a complex number. So, does this mean s has no physical significance at all?

To make sense out of this complex nature of frequency, let us look at the term:

$$e^{st} = e^{(\sigma + j\omega)t} = e^{\sigma t} e^{j\omega t} = e^{\sigma t}(\cos \omega t + j \sin \omega t) \tag{3.62}$$

It is to be noted that we have dropped the negative sign here as we just want to look at the effect of complex frequency and that we are not considering the term in the

Laplace Transform. Euler's equation can be used to express the second exponent $e^{j\omega t}$ as the cosine and sine as done before. However, the first exponent $e^{\sigma t}$ is merely the change in the magnitude with respect to time of the cosine and sine waveforms. The best way to understand this is to plot the waveforms for the three cases—$\sigma = 0$, $\sigma > 0$ and $\sigma < 0$.

Figures 3.5, 3.6 and 3.7 show the function $e^{\sigma t}e^{j\omega t}$ when $\sigma = 0$, when $\sigma < 0$ and when $\sigma > 0$, respectively. When $\sigma = 0$, the resulting sinusoid has a constant magnitude. If $\sigma < 0$, the resulting sinusoid decreases exponentially with time until

Fig. 3.5 $e^{\sigma t}e^{j\omega t}$ when $\sigma = 0$

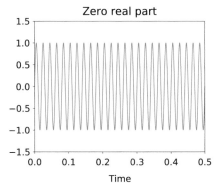

Fig. 3.6 $e^{\sigma t}e^{j\omega t}$ when $\sigma < 0$

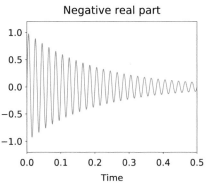

Fig. 3.7 $e^{\sigma t}e^{j\omega t}$ when $\sigma > 0$

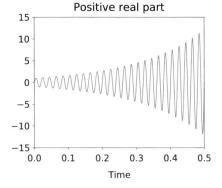

it tends to zero as time tends to ∞. When $\sigma > 0$, the resulting sinusoid increases exponentially with time until it eventually becomes infinite when time tends to ∞. This shows that the real part σ of the complex frequency s merely impacts the magnitude of the signal. The periodic frequency of the signal is dependent on the imaginary part of the frequency $j\omega$.

From the above waveforms, it is fairly clear that the frequency s due to its complex nature affects the stability of the output. If the output increases to becoming unbounded for $\sigma > 0$, the system that produces this output is essentially unstable. This form of stability analysis is very essential in several types of system studies such as control system design and analysis. However, in signal processing, most of our analysis is to understand the periodic oscillatory behaviour of systems. Due to this reason, in signal processing, the real part of s is usually dropped and the frequency s is purely imaginary:

$$s \approx j\omega \tag{3.63}$$

This section discussed the significance of the variable s, which appears in the result of the Laplace Transform. There is usually very little discussion about why this s is the frequency and it is usually taken for granted once we perform frequency response analysis that s is indeed the frequency. However, if step back and look at Euler's equation and how it is used to express sinusoids, it becomes fairly clear how s is the frequency and also how the complex nature of s determines the nature of the output of any given system. With this foundation, let us now look at the advantages of Laplace Transform.

3.9 Advantage of Laplace Transform

Now that we have gone through a fair bit to describe Laplace Transform as an operation and how it produces a function in frequency, let us now look at how it helps. We have seen how Laplace Transform is a transformation that involves an integral operation and that this transformation can be applied to any function in time t. What makes Laplace Transform so popular in signal processing is the effect it has on many of the functions that we use very often.

In signal processing, we always begin with signals as functions in time—for example, a sine or a cosine waveform with respect to time. For example, the voltage available at our home power outlets can be expressed as

$$v(t) = V_{pk} \sin 100\pi t \tag{3.64}$$

Assuming that the frequency of voltage is 50 Hz, which is the case in most of the world except for USA and Canada where it is 60 Hz. This is just one example of a signal. We could think of many more. A signal such as this would be applied as an input to a system. Quite often a system is a term that is dropped vaguely. Let us look at what a system might be.

A system is anything that could receive this input and produce an output that might be different from the input. What is an output? Another signal. And a system could produce several signals from just one input. We could take all of them as outputs or select a few signals that we call output. Let us take another example. Suppose we apply the voltage above to a capacitor C. The current that will flow in the branch will be

$$i(t) = C\frac{dv(t)}{dt} \tag{3.65}$$

This current is a signal that is produced by the capacitor. The capacitor therefore can be called a system that accepts the voltage as an input signal and produces a current as an output signal.

We could connect a number of passive components—inductors, capacitors in several different ways. We could apply the voltage v above to two terminals of the circuit as an input and measure the current flowing in any branch or the voltage across another branch and designate one or more of them as the outputs of this new system. We can write the equations for each component, and if we collect them together, we have equations that describe the system. We could perform substitutions to express the output with respect to the input, and we will have a similar equation for our new system. We have done this already for the LC filter in the past section.

Suppose we perform the Laplace Transform on the differential operation of the capacitor (3.65). We have done this in the previous section, and this turns out to be

$$I(s) = sCV(s) - Cv(0) \tag{3.66}$$

The result is a polynomial in the variable s, which we have seen is the frequency. There is the initial condition $v(0)$. Quite often, in electrical circuits, it is quite normal to consider a system to be initially at rest, which means at time $t = 0$, all variables are zero and so $v(0) = 0$. What is the significance of this conversion of a differential equation to a polynomial equation? Let us look at some examples.

Take a simple polynomial equation:

$$x^2 + 2x + 1 = 0 \tag{3.67}$$

An equation like this can be solved to get its roots—$x = -1, -1$. One wonderful thing about solving it is we know values of x for which it becomes 0. Take another equation:

$$\frac{x - 3}{x^2 + 2x + 1} \tag{3.68}$$

This equation becomes 0 for $x = 3$ and infinite for $x = -1$. And we can plot values for various values of x and find many other properties like slope, maxima, minima etc.

Suppose now we consider a differential equation such as

$$\frac{d^2x}{dt^2} + 2\frac{dx}{dt} + 1 = 0 \tag{3.69}$$

To analyze this, we would have to solve it by integration. The integration would have to be done either analytically or numerically. But, this differential equation is not as convenient from the point of analysis as the polynomial counterpart.

If we consider the example of the LC filter, we had expressed the output voltage with respect to the input voltage by the following double integral:

$$v_o(t) = \frac{1}{LC} \int \int (v_{in}(t) - v_o(t))dt \tag{3.70}$$

The above equation can be transformed using Laplace Transform to produce the following function in frequency s:

$$V_o(s) = \frac{V_{in}(s) - V_o(s)}{LCs^2} \tag{3.71}$$

The latter is way more convenient for analysis as the former simply because it is a polynomial instead of a differential.

In most practical cases, a large number of signals and systems that are functions in time t when transformed using Laplace Transform will result in polynomials in frequency s. There are, of course, complex signals and functions of time that will not result in mere polynomials in s after Laplace Transform and the transformed functions in s may also contain differentials and/or integrals. However, in a majority of cases, practical systems and circuits can be transformed and analyzed as mere polynomials using Laplace Transform.

3.10 Digital Signal Processing

We covered a lot of ground with respect to Laplace Transform. We need to stop and realize something—everything we dealt with so far in this chapter is in continuous time. Even the frequency s that we see in the transformed functions is the continuous time frequency. At first glance, this might not make much sense. We transformed functions from the time domain to the frequency domain using Laplace Transform. Therefore, have we not reached another independent domain where we should not have any notion of time? Unfortunately not. Even frequency has different definitions in continuous time and discrete time.

In reality, frequency is defined rigorously only in the continuous time domain. In the discrete time domain, we need to come up with an alternative definition of frequency. In order to do so, we need to take a step back and have another look at the variable s that we have been calling frequency so far. We had an entire section devoted to how s is indeed the frequency. Now, let us look at how else we can perceive this variable s.

We could think of the variable s in the transformed functions as the differential operator. So, let us revisit the Laplace Transform of the differential of a time varying function:

$$\mathcal{L}\left[\frac{dx(t)}{dt}\right] = sX(s) - x(0) \tag{3.72}$$

This is something we have even derived in the previous section, and we used it to transform the equations for the capacitor and the inductor. However, it is possible to see an equivalence between $\frac{dx(t)}{dt}$ and $sX(s)$ if we drop the initial condition $x(0)$. So, we could think of it as

$$\mathcal{L}\left[\frac{d}{dt}\right] \leftrightarrow s \tag{3.73}$$

We are trying to create an equivalence between the differential operator in the time domain and the variable s in the frequency domain.

In (3.73), we are saying that s, which we have said is a frequency, can also be seen as a differential operator. If s appears in a transformed function, the corresponding time domain function would have a differential at the same spot. This is not strictly mathematically correct as you need to perform the inverse Laplace Transform. However, we need to find a way to apply mathematics and not just rigorously follow it. Therefore, with this relation, we have now projected s as an operator instead of just a frequency.

This new avatar of s takes even more concrete shape if we look at what happens to the Laplace Transform of an integral of a function of time:

$$\mathcal{L}\left[\int x(t)dt\right] = \frac{X(s)}{s} \tag{3.74}$$

which gives exactly the opposite equivalence:

$$\mathcal{L}\left[\int dt\right] \leftrightarrow \frac{1}{s} \tag{3.75}$$

Again, also not strictly mathematically correct, as we cannot apply Laplace Transforms to operators. However, the relation is fairly obvious. So, if a time derivative maps to s and a time integral maps to $\frac{1}{s}$, the position of s in the transformed function can be used to create the equivalent operator in the time domain.

To illustrate this, take the example of the following polynomial, which let us suppose is a function that is the result of a Laplace Transform:

$$\frac{s^2 X(s) + 2s X(s) + 3X(s)}{s} \tag{3.76}$$

If we assume that

$$\mathcal{L}[x(t)] = X(s) \tag{3.77}$$

Equation (3.76) is equivalent to the time domain function:

$$\int \left[\frac{d^2x(t)}{dt^2} + 2\frac{dx(t)}{dt} + 3x(t) \right] dt \tag{3.78}$$

If you perform a Laplace Transform to (3.78) and take the initial condition $x(0) = 0$, you would arrive at (3.76).

The purpose of the above discussion on finding an operator quality to s, which we had initially stated was the frequency, is to try and produce an analogous operator in the discrete time domain. It is quite easy to understand the concept of frequency in the continuous time domain as frequency is also something we can physically define and measure. However, in the discrete time domain, when all we have are samples of a signal, frequency is not that straightforward a definition. What if we assume for simplicity that the frequency in the discrete time domain is the same as that in the continuous domain because all we are doing is sampling a signal and not altering it as a function?

Assuming the frequency stays the same in the discrete time domain however causes a problem. Depending on the rate of sampling, the resultant approximation of a function in the discrete time domain might result in a very different perception of the function. A very high sampling frequency (low sampling time period) would preserve the nature of the function in the discrete time domain quite accurately. However, a low sampling frequency (large sampling time period) could result in the function in the discrete domain being quite different. Since the sampling time plays a critical role in how a signal is represented in the discrete or digital domain, assuming that the frequency remains the same in the digital domain will not be correct.

Instead of being stuck to the definition of frequency, we could think of an alternative definition of s in the discrete time domain. To begin with, it is advisable that we rename the variable when looking at Laplace Transforms in the digital domain. The name of the variable is just a name but by convention, to avoid confusion, in the digital domain, the frequency is denoted by z. The next question, what is this z and how do we relate s with z? This is where the representation of s as an operator helps. If the definition of frequency in the digital domain might be a bit tricky due to sampling frequency, we could think of z as an operator to figure out what it could be.

The first problem we encounter is that in the discrete time domain, we cannot define a differential in the same way. The reason is because to be able to differentiate a signal, the signal needs to be continuous to begin with. A discrete time signal fails that criterion right away because the signal is defined at discrete events. So, first let us look at this in a bit more detail. Continuity of a function (here continuous time) would be best defined by L'Hospital's rule [10] as shown in Fig. 3.8.

In physical terms, continuity of a function at some time t as shown in Fig. 3.8 asks a question—is there a jump in the function at that instant of time? The question is answered by the two limits, one from the left $\lim_{h \to 0^-}$ and the other from the right $\lim_{h \to 0^+}$. If the two limits of the function exist and are equal to each other, the result is there is no jump and the function is continuous at that instant of time. If this rule holds true for every instant of time, the function is said to be continuous everywhere. Such a universally continuous function will also be differentiable. The derivative of a function at any instant of time t is the rate of change of the function at that instant of time t. This derivative will be finite if the function is continuous and has no jump. A universally continuous function is therefore universally differentiable.

Figure 3.9 shows the same function in Fig. 3.8 in the form of samples in the digital domain. It is fairly obvious that with mere samples like these, L'Hospital's rule for determining continuity fails. With the continuity test failing, the differentiability test also fails. So, here is our problem. We cannot define differentiability in the digital domain. So, how are we going to translate s as an operator in the digital domain?

Fig. 3.8 Continuity check for a function

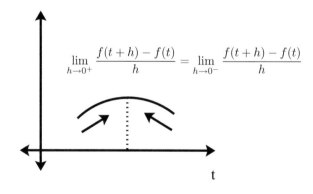

$$\lim_{h \to 0^+} \frac{f(t+h) - f(t)}{h} = \lim_{h \to 0^-} \frac{f(t+h) - f(t)}{h}$$

Fig. 3.9 Lack of continuity in the digital domain

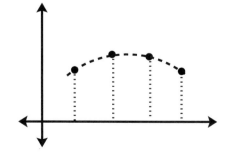

Let us first express the function in the digital domain mathematically. For a continuous function in time $x(t)$, the digital domain representation is written as

$$x[nT] \quad , \text{ for } n = 0, 1, 2, \ldots \tag{3.79}$$

where T is the sampling time interval. The definition is quite simple as it merely states that the continuous function $x(t)$ is defined as a series of samples that are multiples of the sampling time period T. These samples are mere integers n starting at 0 though that is merely a convention to start at time $t = 0$ seconds, but n can be any integer and can be negative as well. It is quite normal to drop the sampling time interval T as it is implicit in a digital domain representation:

$$x[n] \quad , \text{ for } n = 0, 1, 2, \ldots \tag{3.80}$$

We have already seen how continuity and differentiability fail, which means for any particular sample n, we cannot define a derivative at $x[n]$. However, for every sample $x[n]$, we can locate the next sample $x[n + 1]$ that is in the future and we can locate the previous sample $x[n - 1]$. Instead of a differential operator, at the sample n, we define an advance operator and delay operator. Let us start by calling this advance operator z. We will talk about the delay operator soon. We will apply the advance operator to transformed function.

We have seen previously, that for a continuous time function $x(t)$, the Laplace Transform applied to it produces a function in s, which we have by convention decided to call $X(s)$. When we convert the function $X(s)$ to the digital domain, we call it $X(z)$. In the next section, we will discuss how this conversion takes place. For now, just assume that every continuous Laplace Transformed function $X(s)$ can be converted to its digital domain equivalent $X(z)$. We could say that indirectly a digital sampled signal $x[n]$ can be transformed to $X(z)$:

$$\mathcal{L}\{x[n]\} \leftrightarrow X(z) \tag{3.81}$$

In reality, we do not perform Laplace Transform directly on $x[n]$. We perform Laplace Transform on $x(t)$ and convert it to $X(s)$, which is then subsequently converted to $X(z)$.

With the above relation between the digital domain sampled function $x[n]$ and the transformed function $X(z)$, we can now define the advance operator. When the advance operator z is multiplied with the transformed function $X(z)$, this is equivalent to the Laplace Transform of signal samples shifted to the future by one sample:

$$zX(z) \leftrightarrow \mathcal{L}\{x[n + 1]\} \tag{3.82}$$

Therefore, just like $sX(s)$ in the Laplace Transform relates to the derivative $\frac{dx(t)}{dt}$ in the time domain, $zX(z)$ relates to the advance operator applied to $x[n]$ at sample n to produce the next sample $n + 1$, which is $x[n + 1]$. The advance operator z can be applied to the digital function in exactly the same way as the derivative was applied to the continuous function.

Now to talk about the delay operator. Once again, an analogy with the continuous time domain. In the continuous time domain, s was shown to be equivalent to the differential operator $\frac{d}{dt}$, while its reciprocal $\frac{1}{s}$ was shown to be equivalent to the integral operator $\int dt$. We let z be the advance operator and defined it as above. Therefore, we could use the same reciprocal nature to define the delay operator with respect to the advance operator. We denote the delay operator as $\frac{1}{z}$ or z^{-1}. So,

$$z^{-1} X(z) \leftrightarrow \mathcal{L}\{x[n-1]\} \tag{3.83}$$

The delay operator z^{-1} has exactly the opposite effect as the advance operator. When applied to $x[n]$ at the sample n, the delay operator produces the previous sample $x[n-1]$ at $n-1$.

Advance and delay operators can be applied repeatedly just like higher powers of derivatives or multiple integrations.

$$z^{k} X(z) \leftrightarrow \mathcal{L}\{x[n+k]\} \tag{3.84}$$

$$z^{-k} X(z) \leftrightarrow \mathcal{L}\{x[n-k]\} \tag{3.85}$$

In the above, advance or delay operators applied k times will yield k samples to the future $x[n+k]$ or k samples to the past $x[n-k]$, respectively.

We now have defined how the frequency z in the digital domain can be expressed as an operator. In the next section, we will look at how we can convert a Laplace Transformed function from the continuous s domain to the digital z domain.

3.11   Continuous to Digital Conversion

In the previous section, we examined how s the frequency in the continuous Laplace Transform functions could be perceived as the derivative operator. We had then created an analogous perception for z the frequency in the digital or discrete Laplace Transform domain as the advance operator. In this section, we will discuss how we can transform a function from the continuous Laplace Transform domain to the digital Laplace Transform domain. There are a number of ways to convert a Laplace Transformed function from the continuous domain to the digital domain. In this section, we will discuss a few of them and eventually describe the method that is recommended and will be used going forward in this book.

To tackle the problem of conversion from continuous domain to digital domain, we need to ask the reverse question—if we had a signal in the digital domain in the form of mere samples, how would we recreate the continuous domain signal? The simplest way to transform continuous signal systems to discrete and vice versa

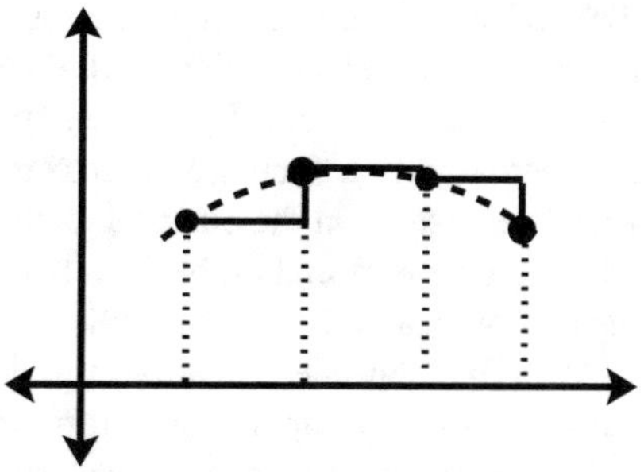

Fig. 3.10 Zero order hold approximation of a sampled signal

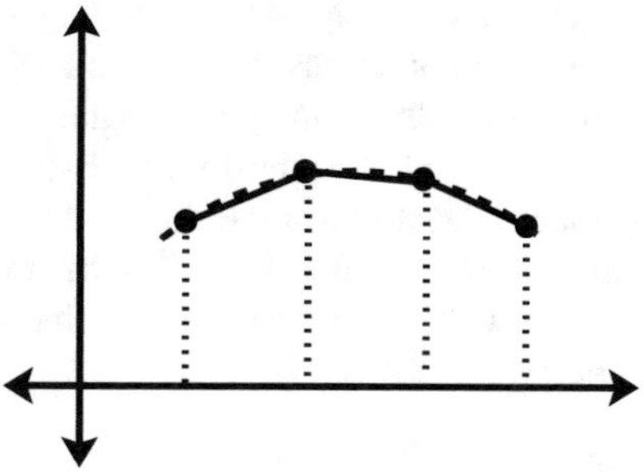

Fig. 3.11 LoopMap for the circuit

is using the zero order hold [1]. Conceptually, zero order hold can be described using Fig. 3.10. In simplest terms, the signal remains constant between two samples and held at the value equal to the most recent sample in the past. This is simplest approximation of a digital signal, and it results in a staircase approximation for the equivalent continuous signal. Though this is fairly simple, it does not completely map a continuous system to a discrete one. More importantly, using this method of conversion between continuous and discrete results in systems that may not be stable.

An improvement over the zero order hold in terms of accuracy and stability is the first order hold [1]. In this case, we approximate the signal between samples as a linear function between the samples. Figure 3.11 shows this concept with respect to a sampled signal. In contrast to Fig. 3.10, we now have a much better approximation since we have straight lines between the samples. Though this might seem like the ideal solution, for more complex functions, the straight line does not capture the potential non-linearity of the signal between two samples. Therefore, the first order hold method also suffers from instability problems for certain types of functions.

The past two methods were briefly described to provide an idea about the approximation that is carried out when converting signals between continuous and digital domains. The method that is most widely recognized to result in accurate and stable approximations is the Bilinear Transformation or the Trapezoidal Rule [1]. Mathematically, the transformation relates z with s as follows:

$$z = e^{sT} \tag{3.86}$$

Physically, the significance of this is to express the advance operator z as a non-linear function (exponential) of the derivative s operator taking the sampling time interval T into account. In contrast, with the zero order hold, the derivative is hard to describe with a staircase approximation due to the discontinuous nature of the waveform itself. On the other hand, with the first order hold, the derivative and the advance operators end up being related by a linear function due to linear nature of the approximation between samples.

With this non-linear relationship, we also now have a complication—if directly substituted, every Laplace Transform in the digital domain will end up being a non-linear function throwing away the major benefit of Laplace Transform. This benefit is to be able to convert time varying functions into polynomials in s. So, we must have a linear relationship between s and z. This produces a conflict. A simple linear function results in an approximation that is not always stable as is the case of the first order hold. So, how do we satisfy both conditions? A linear function that makes Laplace Transform in the digital domain as simple as it is in the continuous time domain. Yet, this linear function has to mirror the non-linear nature of (3.86).

One way to remove the non-linear exponential function is using the infinite series:

$$e^{sT} = \sum_{k=0}^{\infty} \frac{(sT)^k}{k!} = 1 + sT + \frac{(sT)^2}{2!} + \frac{(sT)^3}{3!} + \dots \tag{3.87}$$

Now at least we have removed the exponential and brought in a polynomial. And since it is an infinite series, it is non-linear. However, this does not really work well either as for a reasonably accurate approximation, and we need to consider many terms with several higher powers of sT. Eventually, the Laplace Transform in digital domain will be cumbersome and messy.

We could express (3.86) in another way. Let us begin with this little juggle:

$$z = e^{sT} = \frac{e^{sT/2}}{e^{-sT/2}} \tag{3.88}$$

We have merely split up the exponential into a ratio. We could use the infinite series representation for the numerator and denominator:

$$e^{sT} = \frac{1 + \frac{sT}{2} + \frac{(\frac{sT}{2})^2}{2!} + \frac{(\frac{sT}{2})^3}{3!} + \dots}{1 - \frac{sT}{2} + \frac{(-\frac{sT}{2})^2}{2!} + \frac{(-\frac{sT}{2})^3}{3!} + \dots} \tag{3.89}$$

This may seem like an absolute mess but important to understand what our eventual goal is. This ratio has to mirror the non-linear nature of the exponential or the equivalent infinite series with several higher powers. Since we have a ratio, the division will lead to a series itself. It turns out that even if the equation is simplified to having only first order terms in the numerator and denominator, the resulting division will still be a series that would have sufficient higher order terms.

$$z = e^{sT} = \frac{1 + \frac{sT}{2}}{1 - \frac{sT}{2}} \qquad (3.90)$$

This above equation though is only a ratio of first order polynomials, due to it being a ratio, the equivalent polynomial results in a series that is accurate enough. So, we now have the best of both worlds—a reasonably simple function and at the same time the transformation inherently preserves the complexity of the relationship between z and s. The converse is

$$s = \frac{2}{T} \frac{z - 1}{z + 1} \qquad (3.91)$$

Equations (3.91) and (3.90) are the most accurate mapping between the continuous domain and the digital domain and preserve system stability to the best extent.

We now can convert a Laplace Transform function from the continuous time domain to the digital domain. In the next section, we will describe how all these concepts come together and will be used in the later chapters for analysis and design of filters.

3.12 Putting the Pieces Together

We covered a lot of theory in the past sections. It is necessary to describe how these are connected together and how they can be used. On their own, as described in the previous sections, it can be a bit confusing and not very useful in practice.

To begin with, most physical processes are deterministic, which means that there are mathematical equations that govern them. As these processes progress with time, the mathematical equations typically are functions of time along with other variables as well. In electrical engineering, the behaviour of circuits can also be described by functions of time as we have seen equations of inductors and capacitors being differential equations with respect to time. Therefore, the starting point is usually a time varying equation.

This time varying equation can be transformed using Laplace Transform to a function (in most cases, a polynomial) in a variable s. We have shown that this resultant variable s is a complex frequency. Therefore, the transformed equations now capture how the system behaves as frequency changes as opposed to the original equation, which captures system change with respect to time. This lays the foundation of signal processing where we wish to understand the impact of frequency variation on the system. An example could be of a simple radio where the function is to be able to tune channels by changing the frequency of reception.

The transformed equation in s is a frequency domain representation of the physical process but still in the continuous domain. We can choose a sampling time interval T and convert this to a digital domain representation in a variable z. This z is the frequency in the digital domain but can also be defined as an advance operator. The interpretation of z as an advance operator allows us to represent the function in z with respect to samples of the final system.

The very last step—representation of the function in z with respect to the samples of the final system will be covered in the later chapters when we begin to implement digital filters.

3.13 Conclusions

In this section, we started with basic passive components such as the inductor and the capacitor and examined how the LC filter forms a low pass filter. With respect to the LC filter and its output–input equation being a double integral with respect to time, we felt the need to apply a transformation.

We introduced the Laplace Transform as a way of converting time varying equations and functions to their counterparts in the frequency s. We examined how many commonly encountered signals as well as circuit laws can be transformed using Laplace Transform to polynomials in frequency s. We remarked on the simplicity of the polynomials as well as how s is a complex frequency.

We went ahead to describe how Laplace Transform could be applied in the digital domain. For this purpose, we learned how s can be perceived as an operator instead of only a frequency. We defined analogous operators in the digital domain while also perceiving these operators as frequency z in the digital domain.

We finally examined how we could convert a function in s that is the result of applying Laplace Transform to a function in time to a function in z in the digital domain using the Bilinear Transformation. In the later chapters, we will use both functions in s and functions in z when we analyze and design filters.

References

1. Oppenheim, A. V., & Schafer, R. W. (2001). *Discrete-Time Signal Processing*. Pearson.
2. Rao, B. V., Rajeswari, K. R., Pantulu, P. C., & Rama, K. B. (2012). *Electronic Circuit Analysis*. Pearson Education India.
3. Korn, G. A., & Korn, T. M. (1967). *Mathematical Handbook for Scientists and Engineers* (2nd ed.). McGraw-Hill Companies.
4. Doetsch, G. (1974). *Introduction to the Theory and Application of the Laplace Transformation* (2nd ed.). Springer.
5. van der Pol, B., & Bremmer, H. (1987). *Operational Calculus Based on the Two-Sided Laplace Integral* (3rd ed.). Chelsea Publisher Co.
6. Izadian, A. (2019). *Laplace Transform and Its Application in Circuits*. Springer.
7. Valk, J. A. P. P. (2004). Multi-precision laplace transform inversion. *International Journal for Numerical Methods in Engineering, 60*(5), 979–996, John Wiley.
8. Borwein, J. M., Borwein, P. B., & Dilcher, K. (1989). Pi, euler numbers, and asymptotic expansions. *The American Mathematical Monthly, 96*(8), 681–687, Taylor & Francis.
9. Feynman, R. P. (1977). *The Feynman Lectures on Physics* (Vol. 1). Addison-Wesley.
10. Taylor, A. E. (1952). L'hospital's rule. *The American Mathematical Monthly, 59*(1), 20–24, Taylor & Francis.

Chapter 4
Introduction to Python

4.1 Introduction

In the previous chapters, we covered basic theory with respect to signal processing and digital systems. The later chapters of this book will examine how to design, analyze and implement digital filters using the theory presented. However, in order to make this book a convenient reference material, this chapter will present a basic introduction to NumPy and Matplotlib that will be used in later chapters.

It is important to note that NumPy and Matplotlib are very comprehensive topics on their own, and this chapter will merely provide a brief introduction on NumPy and Matplotlib. It is advisable that the reader has basic familiarity with the Python programming language and some of the basic constructs. There are many reference books for Python programming [1, 2]. Additionally, there are several free and paid tutorials online that can be completed in a few hours to attain a basic level of comfort with Python programming.

This chapter will not cover every aspect of NumPy or Matplotlib but will focus on covering those commands that will be used in the book. The objective of this book is to describe how to analyze and implement digital filters, and therefore, this chapter will ease the reader into the tools necessary for this purpose. An expert in NumPy and Matplotlib could give this chapter a brief glance as the chapter is mainly targeted for those who have not used scientific computing tools with Python before. It is strongly recommended to install Python either through the method described in the beginning of the chapter or by any other method that the reader is comfortable with. The reader is encouraged to execute the code blocks in this chapter and examine the results on their own.

S. V. Iyer, *Digital Filter Design using Python for Power Engineering Applications*,
https://doi.org/10.1007/978-3-030-61860-5_4

4.2   Basic Programming Setup

Before we start programming, this section deals with how to setup a working environment on your computer so that you can code along as you progress through this chapter and the later chapters. All exercises and code samples in this book can be solved using free and open source software. You can, of course, replace free and open source suggestions with proprietary software that you may have a license for. However, to follow along with this book, you have the option to use free (legally free) and open source software only.

All programming will be done using Python and associated packages. Why Python? Python is an easy to use high-level interpreted language that is rapidly gaining in popularity especially for scientific purposes. Along with core Python, we will be using associated packages such as NumPy (Numerical Python extensions), SciPy (Scientific Python) and Matplotlib that are also built using Python. With this array of packages that are all free and open source and can be used with Python programming, we have end-to-end solutions—design, analysis, implementation, simulation and verification.

The first step would be to install Python. There are many ways to install Python and even more so depending on the operating system that you normally use. However, this book will adopt the most beginner-friendly method of installing Python. The simplest way to install a comprehensive list of Python packages is with the Anaconda distribution of Python, which can be found on this link:
https://www.anaconda.com/

On navigating to the downloads section, you can choose to download the installer for your operating system—Windows, Mac OS or Linux. The installer is a very large file (>600 MB). However, Anaconda comes with a large collection of packages that makes a beginner's life much easier. And so, I would suggest that you download this installer.

Installation instructions for Windows, Mac OS and Linux are available in the documentation:
https://docs.anaconda.com/anaconda/install/windows/
https://docs.anaconda.com/anaconda/install/mac-os/
https://docs.anaconda.com/anaconda/install/linux/

Anaconda may need some other packages to be installed prior to launching the installer. So it is advisable to read the instructions for the operating system being used.

It is possible to start using Anaconda directly using your operating system's launch menu. However, I would recommend that anyone aspiring to dive into scientific computing should spend some time getting used to command line usage of Anaconda. In Windows, using your Start menu, look for the Anaconda3 folder and within that look for Anaconda Prompt. In Mac OS or Linux, open any command line terminal. This link contains all the possible commands that you could use with the command line:

https://docs.conda.io/projects/conda/en/latest/user-guide/tasks/manage-environ-ments.html

To get started with Python programming, the most recommended method of setting up a project is to create a new environment. Details of creating an environment can be found in the link above. However, to get you up and going soon, the following command will do:

```
conda create --name dsp python=3 numpy scipy matplotlib
```

The above command will create an environment by the name of "dsp" (feel free to choose another name) by using the argument (−−name). The command also states that this environment should be created with the packages Python (specifically version 3), NumPy, SciPy and Matplotlib. We have specified Python version 3 in which case conda (the Anaconda package installer) will fetch the latest version of Python after 3.0.0. For NumPy, SciPy and Matplotlib, conda will fetch the latest packages available. The reason for specifying python=3 is to ensure that Python version 2 is not installed as it is legacy software but is still available by default on many operating systems.

Now that we have created this environment "dsp", we have created a container in which we have installed the packages we need—NumPy, SciPy and Matplotlib. This container is like a water-tight container where we can maintain packages separately from the host computer's packages or from the packages in other environments. This is the recommended method to start a project to ensure that your project does not corrupt other projects and in turn is not corrupted by other projects. To enter this environment, we need to activate it with this command:

```
conda activate dsp
```

You should see dsp in paranthesis—(dsp) in your command prompt or your terminal. This means the environment is active and we can write programs that use these packages.

4.3 Generating Signals: Creating Numpy Arrays

NumPy arrays are the backbone of almost all scientific computing. Whether for solving equations or for training machine learning models, the basic building block is a matrix. And NumPy gives us an entire battery of tools to create, modify, transform and manipulate matrices. We will use NumPy extensively in this book. The functions available in NumPy are too numerous to cover in this book. We will introduce functions as and how we need them and provide links for extra resources.

To use NumPy, we need to import the package:

import numpy as np

This imports the package numpy and makes it available in our code as the object called np. To begin our journey into learning how to use NumPy, let us pick a task. Our first task is to create and plot a sinusoid. This will not only help us create an array but also to plot it using Matplotlib. To be able to generate a sinusoid, we store the values of the signal at different instants of time in a NumPy array so as to be able to plot them using Matplotlib. There are many ways to do so, but I will describe one method. We need to define the array that contains the time instants. For that, first let us decide what do we want from the signal.

Let us suppose we want 1 s of the waveform. Let us suppose this waveform will have a frequency of 50 Hz, which comes to a time period of 20 ms. Next question is how many samples would be good enough for a smooth waveform? The answer to this question is actually quite vague. Usually, the number of samples good enough to represent any waveform is dependent on the rate at which the waveform changes. Again it is to stress here that we are generating signals for analysis and not have started processing them, which will come later. So, in this first step, it is advisable to generate fairly accurate waveforms to mimic the real world.

Typically, while generating waveforms, I would recommend using a generating frequency of at least 1000 times the highest frequency in a particular signal. Just a rule of thumb and you are welcome to try something else out. So, for our purposes, the generating frequency can be chosen to be 1 MHz.

```
f  =  50.0
T  =  1/f
fg  =  1.0e+6
Tg  =  1/fg
t_duration  =  1.0
```

We are choosing a time duration of 1 s, which implies the duration of the waveform will be for 1 s. This sets up the number of samples needed following our choice of generating frequency.

Let us look at arrays in Numpy. The simplest way to generate an array is using the array method and by passing a Python list to the array function.

```
a  =  np.array([1, 2, 3, 4])
```

This produces an object whose type can be determined by:

```
type(a)
```

```
numpy.ndarray
```

The ndarray is the array object in NumPy and contains the primary data being the array that is being created. The ndarray object also has several attributes and methods associated with the array which are extremely useful.

The array method can be used to generate different types of arrays and is a very useful method. However, if we want to generate an array that is a sequence of numbers, there is another very convenient method called arange. As an example:

```
b  =  np.arange(1, 5)
```

This will produce exactly the same array. The syntax of the arange method is:

```
np.arange([start,] stop [, step])
```

Start is optional and if skipped will be taken as 0, and step is optional and if skipped will be taken as 1.

The major difference from the raw Python range() method is that arange() can take floating point arguments unlike range(). So, it is possible to do:

```
b = np.arange(1, 5, 0.5)
```

will produce:

```
array([1, 1.5, 2, 2.5, 3, 3.5, 4, 4.5])
```

Notice, how the arange() method will not have the stop argument in the array—this upper limit is excluded from the array.

Now that we know how to create an array and more importantly, an array of numbers as a sequence, we can use this method to produce the array of time instants.

```
t_array = np.arange(0, t_duration, Tg)
```

This will produce an array starting with 0 until (but not including) t_duration (which is 1 s) and at a time step of Tg (1 microsecond).

4.4 Array Manipulation

Now that we have seen how to create arrays using NumPy, the next step is to learn how to manipulate them. There are two forms of manipulation—performing computations or applying functions and the other is extracting parts of an array to create another array. Both of these forms of manipulation will be used heavily throughout this book. In this section, we will deal with extracting parts of an array, while the next section will deal with applying functions to an array.

Extraction of a part of an array is called slicing. This concept exists in core Python lists and has been migrated to NumPy arrays, although slicing is much more powerful in NumPy arrays as compared to regular Python lists. Slicing usually is done in a manner similar to accessing an element of an array. Let us consider a sample array:

```
a = np.arange(10)
```

will produce

```
array([0, 1, 2, 3, 4, 5, 6, 7, 8, 9])
```

To access the element at index 3:

```
a[3]
```

will produce 3, as indices in Numpy arrays are counted from 0 just like in core Python lists. If we try to access an element that does not exist, we will get an index out of range error.

We have extracted a single element in the case above. The general way to perform a slice is:

```
<array >[[ start ]: stop [: step ]]
```

Again start is optional and if skipped it is 0. Step is also optional and if skipped it is 1. As an example,

```
a [:3]
```

will produce

```
array([0, 1, 2])
```

The slicing always stops as 1 element before the specified stop in a manner similar to how the arange method works. So stop is index 3, but the slice returns elements until element index $3 - 1 = 2$. The following is an example using start and step:

```
a [2:7:2]
```

will produce

```
array([2, 4, 6])
```

With this background, let us poke around with the time instant array we have just created.

```
t_array [:5]
```

will produce:

```
array([0.e+00, 1.e-06, 2.e-06, 3.e-06, 4.e-06])
```

This shows that the difference between two consecutive samples is 1 microsecond, which corresponds to the generating frequency of 1 MHz.

```
t_array . size ()
```

produces

```
999999
```

size() is a method available with every NumPy array that provides the size of the array.

Though there is a lot more to slicing and extractions with respect to NumPy arrays, we will stop here for now. In the next chapter, the concept of slicing will be used for sampling a signal and applying the digital filter.

4.5   Generating Waveforms: Sine and Cosine Functions

In the past section, we saw how slicing can be used to extract parts of a NumPy array. Another manner of creating an array from an existing array is to apply a function to the array. Every programming language allows the computation of an array with respect to another array. However, due to the fact that NumPy was primarily developed for scientific computation versus other programming languages that were developed for general programming, applying a function to a NumPy array is much simpler and more importantly, much faster.

Let us consider the case of a simple array:

```
a = np.arange(0, 3, 0.5)
```

Suppose we wish to multiply every element of the array by 5. In most other programming languages, we would loop through the array and perform our calculation for every element of the array. We could do the same with this NumPy array as well.

```
x = np.array([5*i for i in a])
```

We are using Python list comprehension to treat the NumPy array as a list and create a new list.

The above code will work. However, when applying a function or a computation to a single element as above, the recommended way of creating the new array is to let NumPy loop through the array

```
x = 5*a
```

In the above case, NumPy figures out that the operation "5*" needs to be performed on every element of the array "a" and loops through the array while performing the operation. The result is the same. However, the reason why this is preferred is because of performance. NumPy's looping method is much faster as compared to looping through a for loop. This is a significant benefit when array sizes are large and the final computations are much faster. Moreover, it is far more convenient to simply write the operation as above, and let NumPy figure out how to apply it over every element of the array rather than write a loop.

This method of creating a new array can also be used with mathematical functions, as an example, the sine function. The sine function is available with the math module available with Python. However, NumPy also has a sine function. The simplest way to use is np.sin. For example:

```
np.sin(30*np.pi/180)
```

I am calculating the sinusoid of 30° by converting degrees into radian by multiplying by $\pi/180$ (π is available as np.pi) as the np.sin method needs the angle in radian. This produces:

```
0.49999999999999994
```

which is really 0.5 except for the resolution of the decimal fraction.

The same can be done for cosine:

```
np.cos(30*np.pi/180)
```

gives

```
0.8660254037844387
```

which is really $\frac{\sqrt{3}}{2}$.

Where NumPy methods score big time is that they can directly handle arrays in a manner similar to the example above. The argument need not be a number that is the angle in radian but can also be an array that has the numbers as radians and the function will applied on all the elements. Let us define an array of angles:

```
a = np.array([0, 30, 45, 90])*np.pi/180
```

Basically, $0°$, $30°$, $45°$ and $90°$ all converted to radians by multiplying the array by $\pi/180$. This form of multiplication of an array by a scalar results in every element of the array multiplied by the scalar as we did above. Applying the sine function:

```
np.sin(a)
```

gives

```
array([0.       , 0.5     , 0.70710678, 1.      ])
```

NumPy has a huge range of mathematical functions—trigonometric, logarithmic, square root, and exponents—that can be applied to arrays. The cosine function can also be applied:

```
np.cos(a)
```

gives

```
array([1.00000000e+00, 8.66025404e-01, 7.07106781e-01,
    6.12323400e-17])
```

This method of applying functions on arrays will be used very often in this book as we generate signals and manipulate them.

4.6 Plotting with Matplotlib

Matplotlib is one of the most popular plotting packages available with Python, and it is also free and open source. In this book, we will be using only Maplotlib for all our plotting purposes. Matplotlib offers a huge range of functions to generate plots as well as design them, many of which we will start with in this section but will expand on later in the book.

To start using Matplotlib, we need to import the pyplot package:

```
import maptplotlib.pyplot as plt
```

Importing maptplotlib.pyplot as plt is merely a convention that is followed in most scientific computing projects but is not necessary. We can now start using plt in our code. Before we jump to plotting the sine and cosine waveforms, let us check out the plotting function first.

Let us generate an array:

```
a = np.arange(10)
```

Fig. 4.1 Plotting a sequence
of integers

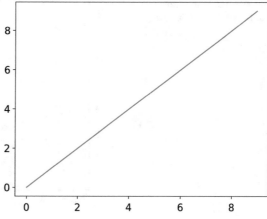

Fig. 4.2 Plotting a multiple
of a sequence of integers

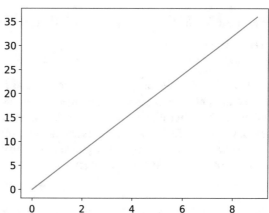

We can plot this by:

```
plt.plot(a)
plt.show()
```

We get a straight line as seen in Fig. 4.1.

A plot like this is fairly useless. Suppose we multiply a by 4 and plot it again.

```
a = a*4
plt.plot(a)
plt.show()
```

Repeating the plot gives a steeper line as shown in Fig. 4.2. It is fairly obvious comparing Figs. 4.1 and 4.2 that the plot function is plotting every element of the array "a" with respect to the index of the element.

We can also specify an x axis rather than just use the index. Let us create another array from "a":

```
x = a*0.1
```

Fig. 4.3 Plotting an array
with respect to another array

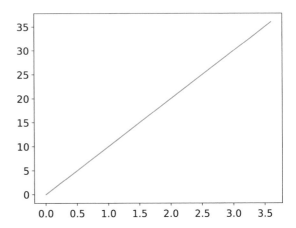

We are artificially creating a time scale as x axis. Now:

```
plt.plot(x, a)
plt.show()
```

produces Fig. 4.3. We are explicitly specifying to the plot command that the array
x (first argument) is the x axis and array a (second argument) is the y axis. The x
and y axis are determined only from positional arguments. The only condition here
is that the arrays for the x and y axis should have the same size or you get an error.

Another useful concept is that you can have two waveforms in a plot. Suppose
we create another array "b" such that:

```
b = 2*a
```

We could plot both "a" and "b" in the same plot as follows:

```
plt.plot(x, a, x, b)
plt.show()
```

produces Fig. 4.4. In the plot command, we are explicitly stating—plot array a on
the y axis with respect to array x on the x axis as well as plot array b on the y axis
with respect to array x on the x axis. It is extremely important to note that the two
waveforms are independent. They do not have to have a common x axis though it
would make little sense to mess around with that.

```
plt.plot(x, a, b, x)
plt.show()
```

produces Fig. 4.5 and is perfectly acceptable as far as Matplotlib is concerned. All
the plot functions need to know that for the first waveform the x axis is array x and
the y axis is array a, while for the second waveform the x axis is array b and the y
axis is array x.

For each waveform, the x and y axis arrays should have the same length.

```
plt.plot(x, a, b, x[1:])
```

Fig. 4.4 Plotting two arrays
in one plot

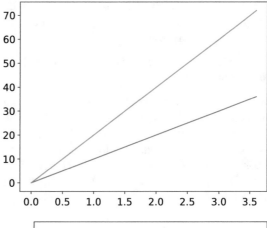

Fig. 4.5 Reversing the plot
axes

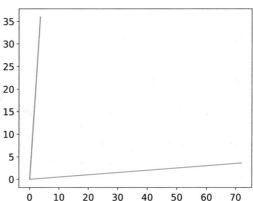

This will not do. The second waveform has incompatible x and y axis array since
we are using slicing to decrease the length of x. The entire command fails. However,
the waveforms need not have the same length between each other. For example:

```
plt.plot(x, a, b[1:], x[1:])
plt.show()
```

will work just fine because for the second waveform, the x axis array and the y
axis array have the same length as the same slicing has been applied. However, the
lengths of the arrays of the first plot being greater than the lengths of the arrays of
the second plot are perfectly acceptable and will not result in an error.

We could use array slicing strategically to extract parts of an array for plotting.
As an example, consider the following arrays:

```
x_short = x[::2]
b_short = b[::2]
```

We are using the step argument of slicing to pick out every alternate element of arrays x and b. The resultant arrays x_short and y_short have half the lengths of x and b, respectively. But:

```
plt.plot(x, a, x_short, b_short)
plt.show()
```

will also work perfectly fine. Because all that counts is that in each waveform the y axis array has to be compatible with its x axis array.

In this section, we examined the basic plotting functionality of Matplotlib. It is important to note that Matplotlib is a very comprehensive package, and in this book we will be using a very small fraction of what Matplotlib can deliver. Data visualization is in itself a vast topic, and the interested reader should explore Matplotlib documentation and tutorials online.

4.7 Plotting Waveforms

In the past couple of sections, we saw how we can generate arrays with the sine and cosine functions applied on every element. We had, for the purpose of illustration, used sample arrays with a few angles. We had also plotted a few sample arrays. In this section, we will bring these two concepts together to create and plot sine and cosine waveforms that we will use later in the book.

In general, for a sinusoid of frequency f, the angle at a time instant t is called the phase angle and is:

$$\theta = 2\pi f t = \omega t \tag{4.1}$$

This means, we can generate the sine and cosine as:

$$\sin \omega t \tag{4.2}$$

$$\cos \omega t \tag{4.3}$$

We have already generated the time instant array in the previous section. We can generate sine and cosine arrays for each time instant with the following block of code.

```
omega = 2*np.pi*f
sin_array = np.sin(omega*t_array)
cos_array = np.cos(omega*t_array)
```

We have defined another variable omega and used np.sin and np.cos functions on the array omega*t_array, which is merely omega multiplied to every element of the time instant array. The entire block code is therefore:

Fig. 4.6 Sine wave

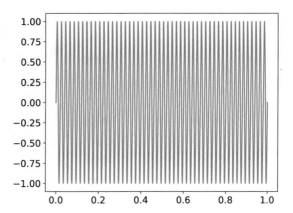

```
f = 50.0
omega = 2*np.pi*f
T = 1/f
fg = 1.0e+6
Tg = 1/fg
t_duration = 1.0
t_array = np.arange(0, t_duration, Tg)
sin_array = np.sin(omega*t_array)
cos_array = np.cos(omega*t_array)
```

With the above sine and cosine arrays, we can now use the plot function in the previous section to plot the waveforms.

```
plt.plot(t_array, sin_array)
plt.show()
```

produces the sine waveform as shown in Fig. 4.6.

```
plt.plot(t_array, cos_array)
plt.show()
```

produces the cosine waveform as shown in Fig. 4.7. We could combine the two plots together in one command to get the plot shown in Fig. 4.8:

```
plt.plot(t_array, sin_array, t_array, cos_array)
plt.show()
```

As is quite obvious from the above plots, a bit of visual design would make them a bit more useful. When using Matplotlib, the above commands will result in a new window that has some interactive functionalities such as zoom where you can expand on a part of the plot. However, in some cases, the end result is to create a plot that can be directly saved to a figure file usually for the purpose of inclusion in a report. There are many commands available to beautify plots. However, a few are usually very common and these will be shown below.

Fig. 4.7 Cosine wave

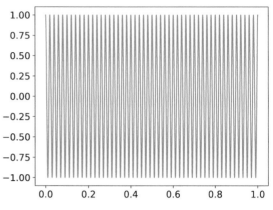

Fig. 4.8 Sine and cosine waves

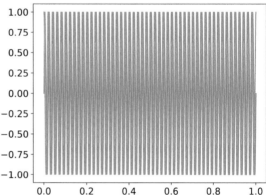

The usual size of the plot is usually a bit small. To specify a dimension of the plot, we can use the figure method with plt.

```
plt.figure(figsize=(10,8))
plt.plot(t_array, sin_array, t_array, cos_array)
plt.show()
```

The figure method to begin with creates a new plot. Until a new figure method is encountered, all plot functions will result in waveforms being added to this plot. The figsize argument helps to configure the plot that is being generated next.

If you notice, the range of the plot is decided by the plot method based on the data. This may not be what you want. Suppose you want the plot to have an x range from 0 to 0.04, you can use the xlim method:

```
plt.figure(figsize=(10,8))
plt.plot(t_array, sin_array, t_array, cos_array)
plt.xlim([0, 0.04])
plt.show()
```

Fig. 4.9 Sine and cosine
waves with a x axis zoom

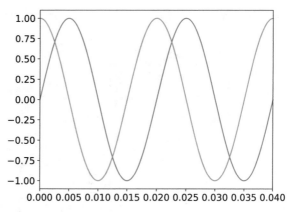

Fig. 4.10 Sine and cosine
waves with a x and y axes
zoom

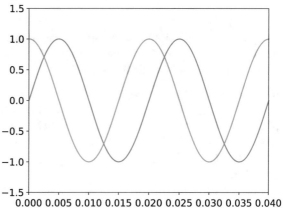

to produce Fig. 4.9. The xlim method accepts as arguments either the start and stop
of the x axis plot or a single argument as a list that has the start and stop as list
elements.

In a similar way, you could also change the y axis range with the ylim method:

```
plt.figure(figsize=(10,8))
plt.plot(t_array, sin_array, t_array, cos_array)
plt.xlim([0, 0.04])
plt.ylim([-1.5, 1.5])
plt.show()
```

Figure 4.10 has both the x and y axes ranges defined to make the plot more clearly
visible.

Since there are only two waveforms in this plot, you can figure out one waveform
from the other in Fig. 4.10. But in case there are many, you will lose track of them
unless you specifically label them and mark the labels somewhere in the plot. In
that case, you need a legend. To add a legend, you need to do a few things. First, we
need to break up our plot statements. As stated before, a figure() method creates a

Fig. 4.11 Plot with a legend

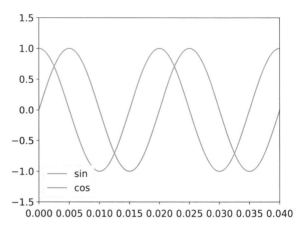

new plot object. All plot functions subsequent to that will result in waveforms that are added to this figure. When the next figure() method is encountered, a new plot is created for the plot functions that follow.

```
plt.figure(figsize=(10, 8))
plt.plot(t_array, sin_array, label='sin')
plt.plot(t_array, cos_array, label='cos')
plt.xlim([0, 0.04])
plt.ylim([-1.5, 1.5])
plt.legend()
plt.show()
```

This results in Fig. 4.11. Each plot function will have a label that can be any string. You must break up the statements when specifying a label because once the label keyword is encountered, additional plot array combinations cannot follow after that. Adding a label to each plot function is not sufficient for the labels to be displayed in the plot figure. In the end, you need to call plt.legend() to add those labels to the legend of the plot.

With the plot now looking much clearer than before, we could add the final touches. We could give our plot a name using the title() method and give the x and y axis names using the xlabel() and ylabel() methods:

```
plt.figure(figsize=(15,10))
plt.plot(t_array, sin_array, label='sin')
plt.plot(t_array, cos_array, label='cos')
plt.xlim(0, 0.04)
plt.ylim(-1.5, 1.5)
plt.legend()
plt.title('cosine and sine')
plt.xlabel('Time')
plt.ylabel('Mag')
```

Fig. 4.12 Plot with names and dimensions

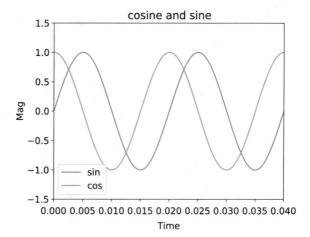

produces the plot of Fig. 4.12.

As already stated, Matplotlib is a very comprehensive plotting package and a detailed description of all the features available is beyond the scope of the book. However, data visualization is a valuable skill and I would encourage the interested reader to look for online tutorials on this topic.

4.8 Conclusions

This chapter serves to provide a basic reference to NumPy and Matplotlib to make the reading in the later chapters a little easier. The chapter only focused on those commands that will be used later to generate signals and plot them. In the later chapters, advanced commands with NumPy will be presented. Furthermore, the SciPy package will also be presented that has the signal package which we will use for signal processing. Commands and functions will be presented gradually as was done in this chapter and as is found necessary.

It is strongly recommended that before moving on to the next chapter, the readers should have a working Python environment on their computers. Furthermore, the readers should be comfortable with generating waveforms and plotting them. The concept of array slicing will be dealt with in greater detail in the next chapter, though it is advisable to be familiar with the concept with the examples given.

References

1. Downey, A. B. (2015). *Think Python: How to Think Like a Computer Scientist* (2nd ed.). O'Reilly.
2. Beazley, D., & Jones, B. K. (2013). *Python Cookbook* (3rd ed.). O'Reilly.

Chapter 5
Implementing Analog Filters Digitally

5.1 Introduction

In Chap. 3, we covered some basic signal processing theory. We introduced the Laplace Transform and examined the advantages of using Laplace Transform particularly in electrical engineering as it converts many of the commonly occurring time varying signals to polynomials in frequency. We also examined how Laplace Transform can be applied in the digital domain and the eventual translation to a time domain equation in terms of signal samples.

In this chapter, we will use the Laplace Transform and the digital conversion process to generate digital models for the commonly occurring components in power engineering. We will begin with passive components such as inductors and capacitors and derive the digital models for them. We will then present Python code to implement these models in a simulation. We will examine simulation results to verify that these models are equivalent to their analog counterparts. Eventually, we will derive the digital model for the LC filter. We will repeat the process of deriving the mathematical model and implementing the digital model with Python code.

The main objective of this chapter is to describe to the reader that most filtering that is performed digitally can be perceived in terms of analog circuits [1, 2]. In many ways, the LC filter forms the basis of most filtering that is implemented in power engineering. This chapter bridges the gap between well-known analog circuits and the digital world of filters. In the next chapter, we will begin to examine filters without any reference to analog circuits. However, before doing so, this chapter will describe to the reader that digital filters under the hood behave very similar to their analog counterparts.

© The Editor(s) (if applicable) and The Author(s), under exclusive license to Springer Nature Switzerland AG 2020
S. V. Iyer, *Digital Filter Design using Python for Power Engineering Applications*,
https://doi.org/10.1007/978-3-030-61860-5_5

Fig. 5.1 Branch with
capacitor

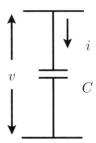

5.2   Implementing a Capacitor Filter Digitally

In this section, we will begin by writing code that will emulate a capacitor. The
first step is to express a capacitor as a mathematical system in contrast to a circuit
component [1, 2]. Therefore, in order to be able to emulate the capacitor, we need
to perceive it as a black box that receives an input and produces an output. In the
previous chapter, we have examined how the behaviour of the capacitor can be
inferred from the equations that describe it. In this section, we will begin with the
equations to derive a black box model for the capacitor.

Figure 5.1 shows a capacitor connected across two nodes. This is typically the
case where a capacitor is used as a filter in a circuit in order to bypass high frequency
currents and produce a smooth voltage across it. As before, the voltage across the
capacitor is

$$v(t) = \frac{1}{C} \int i(t)dt \tag{5.1}$$

We have already seen that we can perform a Laplace Transform on the above
equation to turn this into an equation in frequency s:

$$V(s) = \frac{1}{sC}I(s) \tag{5.2}$$

In the above equations, we are expressing the voltage across the capacitor with
respect to the current in the branch. As stated before, the purpose of connecting
a capacitor is to produce a smooth voltage across the nodes between which the
capacitor is connected. The output voltage $v(t)$ across the capacitor is therefore the
output of the black box for which we are trying to derive a model. From the above
equations, the voltage is determined by the current in the branch. Therefore, the
current in the branch is the input to the model.

The above equations were time domain and frequency domain equations for the
capacitor but in the continuous domain. We can convert this to the digital domain by
using Bilinear Transformation that expresses the continuous frequency s in terms of
the digital frequency z:

$$s = \frac{2}{T} \frac{z-1}{z+1} \tag{5.3}$$

The detailed description of the above derivation can be found in Chap. 3. Substituting the above produces the converted function in the digital domain:

$$V(z) = \frac{1}{C} \frac{T}{2} \frac{z+1}{z-1} I(z) \tag{5.4}$$

With the above equation, we have a model for the capacitor in the digital domain. However, the question arises, how do we implement a filter in a processor from an equation like this? z in the above equation is still a complex frequency in the digital domain. In Chap. 3, we had described how both s and z can be perceived as complex frequencies as well as operators. z is the advance operator and $\frac{1}{z}$ or z^{-1} is the delay operator. Let us rearrange the above equation:

$$2C(z-1)V(z) = T(z+1)I(z) \tag{5.5}$$

As learned before, z being the advance operator, produces a series of samples in the time domain shifted by one sample in the future:

$$zV(z) \leftrightarrow \mathcal{L}\{v[n+1]\} \tag{5.6}$$

Here n is the index of the sample, $n = 0, 1, 2, \ldots$. This follows from our knowledge:

$$V(z) \leftrightarrow \mathcal{L}\{v[n]\} \tag{5.7}$$

where there is an equivalence between a term in the digital Laplace domain and the sampled time varying signal.

Applying the advance operator, the transformed equation in terms of digital samples becomes

$$2C[v(n+1) - v(n)] = T[i(n+1) + i(n)] \tag{5.8}$$

The above equation is called a difference equation analogous to a differential equation in the continuous time domain.

A difference equation such as the one above is solved iteratively. For every sample n, (5.8) holds true. For example, take a particular sample, $n = 10$:

$$2C[v(11) - v(10)] = T[i(11) + i(10)] \tag{5.9}$$

Usually, when we choose a sample, here $n = 10$, we are at that instant of time. We have information of the 10th sample which is our present and we also have information about the past—samples 0 to 9. However, we do not have information of the future—which is the 11th sample. Most importantly, $v(11)$ is unknown because, we do not know what will be the future output since $v(11)$ is the output of the model. Therefore, we need to modify the way we write our difference equation as (5.8) contains samples from the future.

In general, when converting a Laplace Transformed equation in the digital domain into the time domain with discrete samples, the operator z will always result in future samples of signals. To eliminate this problem, we remove any references to z and replace them by z^{-1}. Equation (5.8) can be rewritten by dividing both sides by z:

$$2C(1 - z^{-1})V(z) = (1 + z^{-1})I(z) \qquad (5.10)$$

We have replaced the advance operator z by the delay operator z^{-1}. Let us see what the transformed difference equation in terms of the delay operator is:

$$2C[v(n) - v(n - 1)] = T[i(n) + i(n - 1)] \qquad (5.11)$$

Now, if we take the same 10th sample – $n = 10$:

$$2C[v(10) - v(9)] = T[i(10) + i(9)] \qquad (5.12)$$

The above equation needs the 10th sample (the present instant) and the 9th sample (one sample to the past) of both the input current i and the output voltage v. This equation can now be solved. What do we know? $i(10)$—the present current because it is the input. $i(9)$—the past input which can be stored in memory from the previous iteration. $v(9)$ is also known, because it is the past output and can also be stored in memory from the previous iteration. $v(10)$ is the only unknown because we are yet to calculate it and this is the present output of the model. So

$$v(n) = \frac{T[i(n) + i(n - 1)] + 2Cv(n - 1)}{2C} \qquad (5.13)$$

is the equation we will solve for every sample to obtain the present output sample.

5.3 Coding the Capacitor Filter

In the previous section, we derived a mathematical model for a digital filter emulation of a capacitor. We expressed the model as a difference equation in terms of input and output discrete samples. In this section, we will generate the code for this digital filter model.

Before we code the digital model of the capacitor, we need to lay the foundation for how a digital system is simulated. In the previous chapter, we had generated sine and cosine waveforms using NumPy and plotted them using Matplotlib. Let us use that concept and begin with generating the input signal for the digital model. Let us take the input current i to be a simple sinusoid

```
import numpy as np
import matplotlib.pyplot as plt


Imag = 5.0                    # Amps
f = 50                        # Hertz
omega = 2*np.pi*f             # rad/s
Tg = 1.0e-6           # microseconds
t = np.arange(0, 1.0, 1.0e-6)
i = Imag*np.sin(omega*t)
```

The above block of code produces two NumPy arrays t and i which we can plot:

```
plt.plot(t, i)
plt.xlim([0, 0.06])
plt.title(''Input current'')
plt.xlabel(''Time'')
plt.ylabel(''Amps'')
plt.show()
```

which produces the plot of Fig. 5.2.

The signal in Fig. 5.2 is the current signal that has been generated at a resolution of 1 μs. By choosing a time step resolution of 1 μs, we ensure a fairly accurate signal. However, the time step would need to be chosen with respect to the resolution and the frequency of the signal that needs to be generated. For a signal of 50 Hz, a

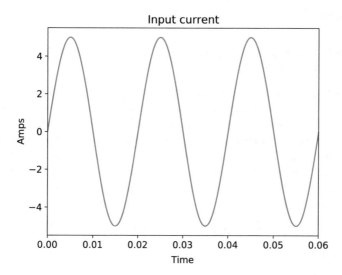

Fig. 5.2 Input current

time step of 1 μs produces an accurate signal. This signal is however the simulated signal. In a practical application, this signal would be obtained from a sensor. As already explained in Chap. 2, the first step in any digital implementation is sampling.

The choice of the sampling frequency depends on our intended application. Most importantly, the sampling needs to be performed such that the resultant digital system captures the signal of interest with sufficient accuracy. To choose the sampling frequency, we use Nyquist criterion. Theoretically, by Nyquist criterion, you need to sample at a frequency of at least twice the frequency of the signal of interest. In practice, though, you need to sample at a frequency of at least ten times the frequency of the signal of interest.

Since, we have a signal of frequency 50 Hz, let us choose a sampling frequency of 5000 Hz. We are therefore sampling at a frequency 100 times greater than the signal. How do we code this? A sampling frequency of 5000 Hz results in a sampling time interval of 200 μs. Since, the input signal is being generated at a time step of 1 μs, this implies the sampling skips 200 signal data elements being generated. Therefore, we can conveniently use NumPy array slicing:

```
T = 200.0e-6   # Sampling time period of 200 mico secs
 or 5 kHz
no_of_samples = int(T/Tg)
t_samples = t[::no_of_samples]
i_samples = i[::no_of_samples]
```

We have created two new NumPy arrays t_samples and i_samples. i_samples is the sampled current that we will be using in the next step for the digital model. The array t_samples will be used for plotting at a later stage but contains the time instants at which sampling is performed. With this, we have now simulated an ADC.

Before we implement the digital filter, let us plot the sampled signal with respect to the original signal:

```
plt.plot(t, i, label=''orig'')
plt.plot(t_samples, i_samples, label=''samp'')
plt.xlim([0, 0.02])
```

The above plot commands produce the plot of Fig. 5.3. Here there appears to be a bug as the two waveforms are overlapping. At first thought, it appears the sampling has failed and the sampled current i_samples and i are the same arrays. However, the reason for the overlap is due to the setting of Matplotlib. When plotting signals in the manner done above, the plot function will interpolate between samples with a straight line. Due to this interpolation, the two plots end up overlapping.

To explicitly differentiate between the sampled waveform and the original simulated signal, we use the "ds" argument in the plot command:

```
plt.plot(t, i, label=''orig'', ds=''steps'')
plt.plot(t_samples, i_samples, label=''samp'',
 ds=''steps'')
plt.xlim([0, 0.02])
```

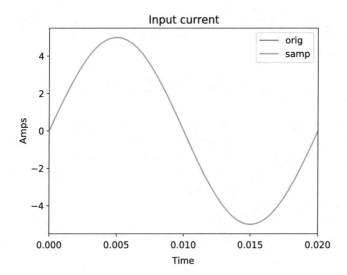

Fig. 5.3 Samples of the input current

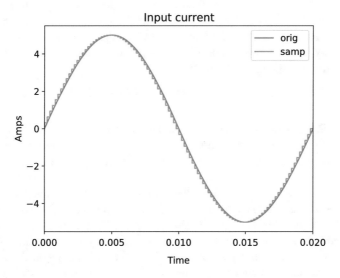

Fig. 5.4 Samples of the input current

Assigning "ds" to "steps" disables the interpolation between data points and now the plot produces a staircase step between two data points. Figure 5.4 shows the sampled current signal as a staircase waveform with respect to the original signal which is still a smooth sine wave. The original signal still appears to be a smooth waveform even though "ds = steps" has been passed to it is due to the resolution of the waveform since we are using a time step of 1 µs versus a sampling time interval of 200 µs.

Since we have extracted samples from the input current, to implement our digital model of (5.13), we need to loop through these samples. Let us revisit this model to plan our implementation:

$$v(n) = \frac{T[i(n) + i(n-1)] + 2Cv(n-1)}{2C} \qquad (5.14)$$

For each sample, we are using present and past samples of both the input and output. Therefore, in this case, conveniently applying a function to the NumPy array will not be possible as we did when we were generating a sine wave. Looping through the samples will allow us the flexibility to write custom code. Moreover, looping through the samples will result in a digital model implementation that can be translated to hardware.

There are two ways to loop through a NumPy array. The first is using the np.nditer method which returns the elements of the array as an iterable. For example:

```
for i_val in np.nditer(i_samples):
        # code
```

With the above for loop, each element of the i_samples array will be available as the variable i_val for the block of code within the for loop. Such a for loop is sufficient for many cases. However, there is another way of looping through a NumPy array which provides not only every element, but also the index of every element. This uses the np.ndenumerate method:

```
for i_index, i_val in np.ndenumerate(i_samples):
    # code
```

The np.ndenumerate method creates an iterable from the array returning a tuple with the index as well as the element. In this book, we will use this method of looping through a NumPy array as it provides the index as well which is useful data.

With the method of sampling established and a way to run computations for each sample, the next step is to begin implementing the model. The first step in the model is to determine the variables and the storage needed. In the model of (5.13), there is the input current i and the output voltage v. In general, when we implement any digital filter, we designate the input to the filter as u and the output from the filter as y. The next step is to allocate memory for these variables. From (5.13), for each calculation of the output, we need the present input, the immediate past sample of the input, and the immediate past sample of the output. For both the input and the output, we need two samples—one for the present signal sample and one for the signal sample in the immediate past. Therefore, the following array initializations will be sufficient:

```
u = np.array([0.0, 0.0])
y = np.array([0.0, 0.0])
```

We have declared the input u and the output y to be arrays with two elements and we have initialized the elements to 0. We could assign one of the elements as the present value and the other element as the immediate past sample. For example, u[0] could be $u[n]$, while u[1] could be $u[n-1]$, while y[0] could be $y[n]$, while y[1] could be $y[n-1]$. Within the for loop, the filter can be implemented as follows:

```
u[0] = i_val
y[0] = ( T*u[0] + T*u[1] + 2*C*y[1] ) / (2*C)
```

We have translated the model of the capacitor to a line of Python code. The present input u[0] will be the current element i_val that is being iterated. The only step that remains is storing the input and output samples. We have initialized u and y to be arrays of 0. Which means, at the very first iteration, u[1] and y[1] will be 0. This is reasonable as we are assuming a system at rest. After performing the above computation, we need to store the present value of input and output.

```
u[1] = u[0]
y[1] = y[0]
```

In a hardware application, the signal sample y[0] which denotes $y[n]$ will be used further in a control loop or for any other computation. In our case, we need to store these output samples in an array so that we can plot the output as a waveform. For this purpose, we need to define an output array. One way to define and create an output array is to initialize it with the np.zeros method.

```
v_samples = np.zeros(i_samples.size)
```

The np.zeros method will create an array of the same size as i_samples. This is provided by the size attribute of a NumPy array. Every NumPy array has a size attribute that returns the number of elements in the array. The zeros method accepts an integer argument and creates a NumPy array with that many elements and initializes all of them to 0.

After each computation, we can assign the output of every sample to the v_samples array.

```
v_samples[i_index] = y[0]
```

This is where the index i_index comes useful. We can assign a particular element of the output array to the present value of the output of the model.

We can now list the entire code for the capacitor as follows:

```
import numpy as np
import matplotlib.pyplot as plt

C = 10.0e-6             # Farad
Imag = 5.0              # Amps
f = 50                  # Hertz
omega = 2*np.pi*f       # rad/s
Tg = 1.0e-6             # microseconds
t = np.arange(0, 1.0, Tg)
i = Imag*np.sin(omega*t)
```

```
T = 200.0e−6                  # Sampling time period of
                              # 200 mico secs or 5 kHz
no_of_samples = int(T/Tg)
t_samples = t[::no_of_samples]
i_samples = i[::no_of_samples]

v_samples = np.zeros(i_samples.size)

u = np.zeros(2)
y = np.zeros(2)

for i_index, i_val in np.ndenumerate(i_samples):
    u[0] = i_val
    y[0] = ( T*u[0] + T*u[1] + 2*C*y[1] ) / (2*C)
    u[1] = u[0]
    y[1] = y[0]
    v_samples[i_index] = y[0]

plt.plot(t, i, label=''orig'')
plt.plot(t_samples, i_samples, label=''samp'')
plt.xlim([0, 0.02])
plt.title(''Input current'')
plt.xlabel(''Time'')
plt.ylabel(''Amps'')
plt.legend()

plt.figure()
plt.plot(t_samples, i_samples, label=''input'')
plt.plot(t_samples, v_samples, label=''output'')
plt.title(''Input and output'')
plt.xlabel(''Time'')
plt.ylabel(''Amps and Volts'')
plt.legend()

plt.show()
```

We have chosen the value of the capacitor to be $10\,\mu F$. We are creating two plots. The first one is the current signal and the sampled input to the digital model. This plot is the same as in Fig. 5.4. With the plt.figure() method we are generating a new plot to which we are plotting the input current samples and the output voltage samples with respect to the time samples.

Figure 5.5 shows the input current and the output voltage of the capacitor digital model. Figure 5.6 shows the zoomed in plot of the output of the capacitor digital model. A quick comment about the magnitude of the waveforms. The peak-to-peak value of the output voltage is expected to be

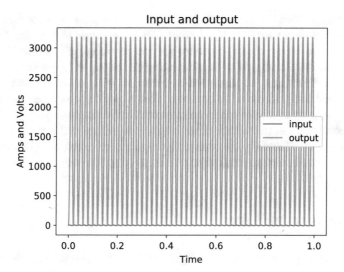

Fig. 5.5 Input and output of the filter

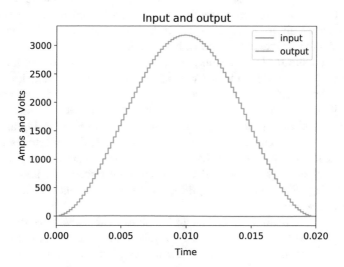

Fig. 5.6 Zoomed in plot of the capacitor model output

$$V_{p-p} = 2 \times \frac{5}{2\pi \times 50 \times 10 \times 10^{-6}} = 3183.09 \text{ V} \qquad (5.15)$$

Therefore, the extreme large peak of the voltage waveform is expected. However, there is one aspect that seems troubling. The output voltage waveform appears to have a dc offset even though the input is a pure ac. The answer to this riddle will be provided in a few sections.

To conclude this section, we have implemented the digital model of the capacitor as Python code and verified it by plotting it. It is important to note that the actual implementation of the digital capacitor are the few lines of code within the for loop. One of the advantages of implementing the digital model within such a for loop is that it can now be translated to hardware with minimal changes. When implementing a digital model in hardware such as a microcontroller, the model is usually repeatedly executed within an Interrupt Service Routine (ISR). This ISR is usually configured to run with a timer interrupt to ensure that sampling occurs at regular intervals. Therefore, the code we have generated in this section is already in exactly the same environment as would be required in hardware.

5.4   The Inductor Filter

In the past couple of sections, we had generated the digital model for the capacitor and had implemented it using Python code. Our eventual objective is to be able to implement a LC filter digitally. Therefore, an intermediate step towards that goal would be to generate a digital model for an inductor and implement it in exactly the same manner we did for a capacitor [1, 2]. In this section, we will begin with the digital model for the inductor as shown in Fig. 5.7.

Just like we did for the capacitor, we need to decide what will be the input to the inductor digital model and what will be the output from the model. As already stated in Chap. 3, inductors and capacitors play very different roles in a circuit as far as their use as filters is concerned. The prime feature of the inductor is that it opposes any change in the current passing through it by producing an induced emf that opposes the change in current. For this reason, the inductor is usually a series element to a voltage source that may contain large high frequency harmonics. Therefore, even though the applied voltage has high frequency harmonics and is rapidly varying with respect to time, the current through the inductor will be much smoother as a waveform.

Let us begin to quantify these statements above with some equations and later some plots. However, as a first step, it is quite easy to see that the input to the inductor model is the voltage v across the inductor while the output of the model

Fig. 5.7 An inductor

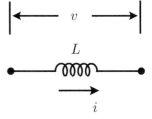

is the current i through the inductor. This is because the inductor is used as a filter producing a relatively smoother current for a particular voltage applied across it.

Let us revisit the equation for the inductor which expresses the current through it with respect to the voltage across it:

$$i(t) = \frac{1}{L} \int v(t) dt \tag{5.16}$$

And, we have seen in Chap. 3 that we can perform a Laplace Transform to turn this into

$$I(s) = \frac{1}{sL} V(s) \tag{5.17}$$

The above continuous domain frequency expression can be converted to the digital domain using the Bilinear Transformation of (5.4):

$$I(z) = \frac{1}{L\frac{2}{T}\frac{z-1}{z+1}} V(z) = \frac{T}{2L}\frac{z+1}{z-1} V(z) \tag{5.18}$$

To be able to create an implementable model for the above digital domain frequency representation, we follow exactly the same steps as we did for the capacitor. We interpret z as an operator besides being also the digital domain frequency. Rearranging the above expression:

$$2L(z-1)I(z) = T(z+1)V(z) \tag{5.19}$$

To avoid the problem of needing future samples of a signal due to the advance operator z in the above equation, we can divide both sides by z to produce

$$2L(1-z^{-1})I(z) = T(1+z^{-1})V(z) \tag{5.20}$$

The above equation can be translated into a time domain expression in terms of signal samples as we did with the capacitor. We use our knowledge of

$$z^{-1}V(z) \leftrightarrow \mathcal{L}\{v[n-1]\} \tag{5.21}$$

$$z^{-1}I(z) \leftrightarrow \mathcal{L}\{i[n-1]\} \tag{5.22}$$

Resulting in the following time domain expression:

$$2L[i(n) - i(n-1)] = T[v(n) + v(n-1)] \tag{5.23}$$

The above expression is true for every sample nT where $n = 0, 1, 2, \dots$.

As already discussed in the previous sections with the capacitor, such an equation can be solved iteratively. We could calculate the present value of the output current $i[n]$ using the present value of the input voltage $v[n]$ and the past samples of current $i[n-1]$ and voltage $v[n-1]$.

$$i(n) = \frac{T[v(n) + v(n-1)] + 2Lv(n-1)}{2L} \tag{5.24}$$

The above digital model of the inductor is very similar to the digital model of the capacitor (5.13). This is in general the form of any typical integral relationship between an input and an output. In the next section, we will present the code for the above mathematical model of (5.24).

5.5 Coding the Inductor Filter

In this section, we will present the code for implementing the digital model of the inductor of (5.24). Since, this is very similar to the implementation of the digital model of the capacitor that has already been covered, a lot of this section will be a repetition.

Just like the case of the capacitor implementation, we will use a generation time step of 1 μs. Since the input to the digital model will be the voltage, we are choosing a 240 V (RMS), 50 Hz sine wave for the input.

```
import numpy as np
import matplotlib.pyplot as plt

L = 1.0e-3              # Henry
Vmag = np.sqrt(2)*240  # Volts
f = 50                    # Hertz
omega = 2*np.pi*f         # rad/s
Tg = 1.0e-6             # microseconds
t = np.arange(0, 1.0, Tg)
v = Vmag*np.sin(omega*t)
```

We have chosen a 1 millihenry inductor as an example. We could plot this signal as follows to get the plot of Fig. 5.8:

```
plt.plot(t, v, ds="steps")
plt.xlim([0, 0.06])
plt.title("Input voltage")
plt.xlabel("Time")
plt.ylabel("Volts")
plt.show()
```

We could sample this input voltage at 5 kHz which results in a sampling time interval of 200 μs. We could generate a v_samples array using NumPy array slicing:

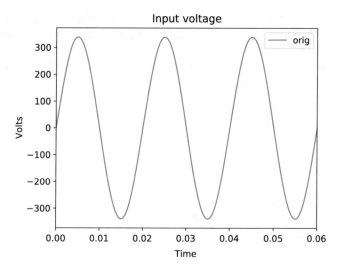

Fig. 5.8 Input voltage

```
T = 200.0e−6    # Sampling time period of 200 mico
    secs or 5 kHz
no_of_samples = int(T/Tg)
t_samples = t[::no_of_samples]
v_samples = v[::no_of_samples]
```

We need to set up our digital filter input and output as well as create an output current NumPy array that we can finally plot.

```
i_samples = np.zeros(v_samples.size)
u = np.zeros(2)
y = np.zeros(2)
```

We initialize the output array i_samples to be an array of zeros having the same size as the input voltage sample array. Since, we will be implementing the digital model of (5.24), we need only one past sample of the input voltage and the output current. The input u and the output y of the digital model can be arrays of two elements each with initial values of 0.

Once we have the input voltage sample array, we can loop through it and implement our model for each element thereby calculating every sample of the output current:

```
for v_index, v_val in np.ndenumerate(v_samples):
    u[0] = v_val
    y[0] = ( T*u[0] + T*u[1] + 2*L*y[1] ) / (2*L)
    u[1] = u[0]
    y[1] = y[0]
    i_samples[v_index] = y[0]
```

The implementation is very similar to that of the capacitor. We are now looping through the voltage array and assigning the output of the model to the current sample array.

The entire code for the implementation of the digital model of the inductor is as follows:

```
import numpy as np
import matplotlib.pyplot as plt

L = 1.0e-3                  # Henry
Vmag = np.sqrt(2)*240       # Volts
f = 50                          # Hertz
omega = 2*np.pi*f               # rad/s
Tg = 1.0e-6                 # microseconds
t = np.arange(0, 1.0, Tg)
v = Vmag*np.sin(omega*t)


T = 200.0e-6    # Sampling  time  period  of  200  mico  secs
    or 5 kHz
no_of_samples = int(T/Tg)
t_samples = t[::no_of_samples]
v_samples = v[::no_of_samples]
i_samples = np.zeros(v_samples.size)
u = np.zeros(2)
y = np.zeros(2)


for v_index, v_val in np.ndenumerate(v_samples):
    u[0] = v_val
    y[0] = ( T*u[0] + T*u[1] + 2*L*y[1] ) / (2*L)
    u[1] = u[0]
    y[1] = y[0]
    i_samples[v_index] = y[0]

plt.plot(t, v, label=''orig'', ds=''steps'')
plt.plot(t_samples, v_samples, label=''samp'',
  ds=''steps'')
plt.xlim([0, 0.02])
plt.title(''Input voltage'')
plt.xlabel(''Time'')
plt.ylabel(''Volts'')
plt.legend()

plt.figure()
plt.plot(t_samples, v_samples, label=''input'',
  ds=''steps'')
```

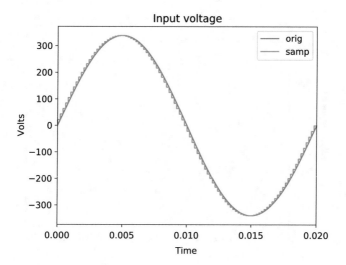

Fig. 5.9 Sampled input voltage

```
plt.plot(t_samples, i_samples, label=''output'',
  ds=''steps'')
plt.xlim([0, 0.02])
plt.title(''Input and output'')
plt.xlabel(''Time'')
plt.ylabel(''Volts and Amps'')
plt.legend()

plt.show()
```

The results are the plots of Figs. 5.9, 5.10 and 5.11.

The magnitude of the output current can be explained by the equation:

$$I_{p-p} = 2 \times \frac{240\sqrt{2}}{2\pi \times 50 \times 10^{-3}} = 2160.75 \text{A} \tag{5.25}$$

As with the case of the capacitor, the riddle that remains is the presence of the dc offset. This will be explained in the next section.

In this section, we have presented the code and the simulation results of the digital model implementation of an inductor. In the next few sections, we will bring in the concept of inductors and capacitors that have losses. Before that, we will examine the need for including the loss in the digital models.

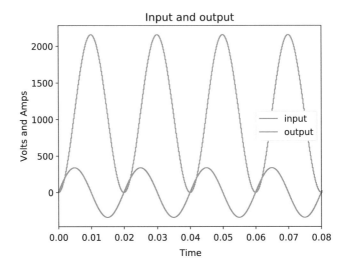

Fig. 5.10 Input voltage and output current

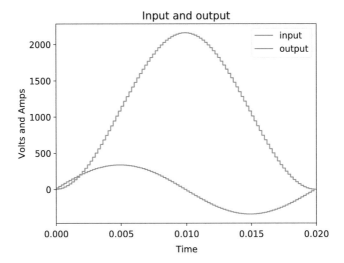

Fig. 5.11 Zoomed input versus output

5.6 The dc Offset

In the implementations of the digital model of the capacitor and the inductor, we found that the outputs had dc offsets. First question is—where did this dc offset come from when we are providing pure ac sine waveforms as input to the models? Is it a bug in the code? In this section, we will examine the reason for this dc offset.

The answer is no, this is not a bug. It is the nature of integration as an operation to produce an offset. Indefinite integration (without upper and lower limits) always produces an integration constant. For example,

$$\int \sin t \ dt = -\cos t + A \tag{5.26}$$

where A is the integration constant or integration offset.

Let us look at what happens when we perform the operation of integration on a sine wave. Numerically, integration is just the area under the curve. We take the product value of the function at a certain time instant and the integration time step. This product is added to a sum which is the result of the integration. The question arises, what will be the initial value of the sum or the integration?

In the case of the integration of the sine wave above, if we know beforehand the exact function the input wave is, we could choose the initial value of the integration to avoid an integration offset. For example, if

$$f(t) = \int \sin t \ dt \tag{5.27}$$

Theoretically, we know the integral of a sine wave to be cosine wave with a negative sign:

$$f(t) = -\cos t \tag{5.28}$$

To eliminate the integration time constant, we would need to ensure that the initial value (at time $t = 0$) of this function $f(t)$ is

$$f(0) = -1 \tag{5.29}$$

Here we are defining the output function $f(t)$ with a non-zero initial value (at time $t = 0$). This was possible because we knew that we were integrating a sine wave $\sin t$. However, in practical cases, when we integrate a signal in real time, we do not know the nature of the input signal. Therefore, we cannot precisely choose a function for the output signal such that an integration offset can be eliminated completely. On the contrary, it is quite normal to start the integration with an initial value of 0. In our implementation of the digital model of the capacitor and the inductor, the integration follows the initial value of the input at time $t = 0$. We have considered the inputs to both models as sine waves which have values of 0 at time $t = 0$. At the same time, the integration output begins at 0 as well.

If we choose an initial value of the integration to be 0 i.e. $f(0) = 0$:

$$f(0) = -\cos 0 + A = 0 \tag{5.30}$$

We end up with a non-zero integration offset $A = 1$. Therefore, the complete equation of the integration is

$$f(t) = \int \sin t \, dt = -\cos t + 1 \tag{5.31}$$

One quick look at the above equation and it is quite obvious that the integral will have a dc offset. In the above case, the entire waveform is above 0 since the lowest possible value of $\cos t$ is -1 and therefore, the minimum of $f(t)$ is 0. From the simulation results of the past two sections, this is exactly what we have got—a waveform with a dc offset. Therefore, the simulation results of the capacitor and the inductor are not incorrect and a dc offset is expected.

5.7   Do We Have Offsets in Reality?

In the previous section, we saw how the dc offsets we observed in the simulations of the digital capacitor and inductor were expected given the mathematical model we were trying to solve. So, with this confirmation that our code is not incorrect, the next question is how is this possible? This does not happen in practice. If you apply an ac sine wave voltage across an inductor, the current through an inductor does not have a dc offset. So how do we get our digital models to mimic what happens in practice?

To begin with, our knowledge of what happens in practice is based on steady state conditions. When a pure ac sine wave is applied as a voltage across an inductor, in steady state, the current flowing through the inductor will also be a pure ac wave with no dc offset. And what happens before the steady state? Dc offsets are possible. The greatest example of these kind of transients arises with machines such as transformer and motors. When machines are started direct online which implies that any random ac voltage is applied across their terminals, these machines draw very large inrush currents that are known to have dc offsets. Depending on the rating of the machine, these transients can take several cycles to disappear.

Our expectation that applying a pure ac sine wave as an input to an inductor or a capacitor will produce an output that is also pure ac is therefore incorrect. Depending on the nature of the element or component being modeled digitally, it is possible that dc offsets may appear either temporarily or even permanently. In the case of the inductor or capacitor, we have seen from their mathematical equations that dc offsets are expected. Let us first examine what this means from a physical standpoint [3].

Let us consider an inductor. Initially, suppose the current flowing through the inductor is zero. In this condition, the magnetic flux in the inductor core will also be zero initially. Let us apply a pure sine wave voltage across the inductor. Let us examine step by step how the current will change. The magnitude of the induced emf is determined by Faraday's Law:

$$e \propto N \frac{d\phi}{dt} \tag{5.32}$$

In the above equation N is the number of turns of the inductor coil, ϕ is the flux and e is the induced emf.

The induced emf is dependent on the rate of change of flux linked with the coil. Therefore, a reverse relationship can also be constructed. The flux can be expressed with respect to the induced emf:

$$\phi \propto \int e \, dt \tag{5.33}$$

The number of turns has been dropped for simplicity. We can expand on the induced emf:

$$\phi \propto \int (v - iR) dt \tag{5.34}$$

where R is the parasitic winding resistance of the inductor coil, v is the external applied voltage and i is the current through the inductor.

We are controlling the external voltage v. To determine how the current through the inductor behaves, let us go back to the fundamental property of the inductor. The inductor will not allow the current through it to change instantaneously as the current and the flux are linked in a proportional relationship:

$$\phi \propto i \tag{5.35}$$

The magnetic flux in turn cannot drastically change as the magnetic field builds up as energy stored in the field gradually increases.

This leads us to the following approximate relationship at the time instant $t = 0^+$ when the external voltage is applied:

$$\phi \propto \int v \, dt \tag{5.36}$$

With this, we are back to our integral relationship. So, if we neglect the integration offset and the applied voltage was a sine wave, the flux ϕ would be a negative cosine wave.

This leads us to a contradiction. If the flux ϕ is a negative cosine wave in this particular case, that implies it will jump to a non-zero negative maximum. This is a direct violation of what we have stated above. The flux cannot change drastically but builds up gradually as stored energy in the magnetic field increases. The only answer lies in including the integration offset:

$$\phi \propto \phi_{ac} + \phi_0 \tag{5.37}$$

The above equation makes it evident how the dc offset in the current can be visualized in terms of the dc offset ϕ_0 in the magnetic field. This dc offset will appear to ensure that the flux in the core is 0 initially because of the property of the inductor by which the magnetic field in its core builds up gradually. A very similar reasoning can be used for the capacitor except that in the case of a capacitor, there is an electric field.

Now that we have described the physical significance of the dc offsets, the next question is why is it that they disappear in steady state? The inductor that we have simulated in the previous sections never had a winding resistance. In (5.34), the flux does take into account the winding resistance of the inductor. This winding resistance represents the loss of the inductor. The loss of the inductor can be a combination of losses—the simple ohmic losses in the winding, magnetic core losses such as hysteresis and eddy current losses and any other losses. Eventually the dc offset that appears in the flux as well as the current will be dissipated by the losses. The same argument also applies to the capacitor. There are losses in the capacitor—terminal ohmic losses and electric field losses.

This section described the nature of offsets in inductors and capacitors. To best understand how they appear and eventually decay, in the next sections, we will model capacitor and inductors with their losses and simulate their digital models.

5.8 Lossy Capacitor

In the previous section, we examined the origin of dc offsets in the inductor and capacitor. In this section, we will generate a model for a lossy capacitor and simulate it [3]. The simulation will make it clear how the dc offset decays with time resulting in a steady state waveform that is pure ac.

To begin with the inclusion of loss in the digital model of the capacitor, we go back to the basics of the capacitor. Figure 5.12 shows a capacitor which is two metallic terminals separated by a di-electric material that prevents the flow of electric current through it but allows the lines of force of an electric field to pass through. When an external voltage is applied across the terminals of the capacitor, charge accumulates on the terminals and an electric field forms across the di-electric as shown. The phenomenon is far more complex than this brief description, but for our objective of generating a mathematical model, we do not need to go very deep.

We wish to develop a mathematical model for the capacitor that includes the losses. In Fig. 5.12, there is no loss and the capacitor is merely the detailed version of Fig. 5.1. Let us define the losses that we wish to model. There is an ohmic loss associated with the terminals of the capacitor as well as the leads to the capacitor. This loss is the ohmic loss which needs only a parasitic resistance to be modeled. However, the loss that we are interested in is the loss associated with the electric field. The electric field is produced in the di-electric of the capacitor when the terminals of the capacitor accumulate charge. The charge on the capacitor terminals and the voltage across the terminals are related through the capacitance:

Fig. 5.12 Input voltage and
output current

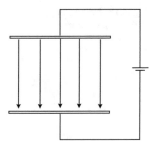

$$Q = CV \qquad (5.38)$$

Here it is important to differentiate between the external voltage source connected across the capacitor terminals and the voltage across the capacitor terminals.
These two quantities are not the same. Initially, when the capacitor is uncharged,
the voltage across the terminals is zero. Applying an external voltage source will
cause a current i to flow that is limited by the impedance external to the capacitor.
As current flows, the charge on the capacitor terminals increases:

$$Q = \int i\,dt \qquad (5.39)$$

As the charge on the capacitor terminals accumulates, the voltage across the
capacitor terminals increases. With increase in the voltage across the terminals, the
strength of the electric field through the di-electric increases. The strength of the
electric field is directly proportional to the voltage applied across the terminals.

$$\overline{\mathcal{E}} \propto v = \frac{1}{C} \int i\,dt \qquad (5.40)$$

Other factors that affect the electric field strength are the construction of the
capacitor—distance between the terminals and the nature of the di-electric medium.
Therefore, as long as a voltage is applied across the terminals, an electric field will
be established across it that is proportional to it.

The voltage v being the output of our digital model was having a dc offset due to
the integration performed on the input current i. We have already examined how the
integration results in an integration offset. Physically, this integration offset ensures
that the output voltage (and the resultant electric field) gradually increases from
zero. In a capacitor, the electric field cannot change drastically and can only change
as charge accumulates on the terminals. The problem encountered in the previous
simulation result was that the integration offset that resulted persisted and never
decayed.

In Fig. 5.12, we have considered a pure capacitor without any loss. Once charge
accumulates on the capacitor, it will remain as it is until the external voltage
changes. Moreover, if the external voltage were to be disconnected and the capacitor

Fig. 5.13 Lossy capacitor
with parallel resistor

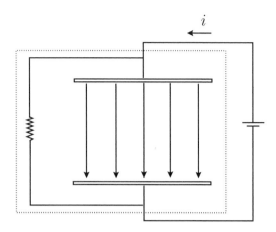

left open-circuited, the charge on the terminals will be retained indefinitely. This, quite obviously, is not the case in practice. The charge on a capacitor is known to "leak". In technical terms, the capacitor has a loss associated with the charge that results in a continuous loss of charge with respect to time. When a capacitor is charged by an external dc voltage source, it will continuously lose charge due to this loss component. Therefore, current will flow continuously from the external voltage source to replenish the charge that has been lost. If the capacitor is disconnected from the external source, the charge will eventually decay to zero leaving the capacitor completely uncharged.

In order to model this loss component, we need to connect a resistor. This resistor must dissipate loss as long as the capacitor terminals have charge. As we have already seen before, if the capacitor terminals have charge, a voltage is present across the terminals and therefore, an electric field is present in the di-electric. In order to dissipate a loss as long as the voltage across the capacitor in non-zero, we need to connect a resistance in parallel with the capacitor and across the terminals as shown in Fig. 5.13.

From Fig. 5.13, the dotted line indicates the new boundary of the capacitor model. The resistor connected in parallel is internal to the model and results in a current $\frac{v}{R}$ as along as there is a voltage across the capacitor terminals. With this addition, the voltage across the capacitor can be expressed as

$$v = \frac{1}{C} \int \left(i - \frac{v}{R} \right) dt \tag{5.41}$$

To create a digital model, we follow the usual steps. First, we perform Laplace Transform:

$$V(s) = \frac{1}{sC} \left(I(s) - \frac{V(s)}{R} \right) \tag{5.42}$$

Converting this continuous domain frequency equation to digital domain with (5.4):

$$V(z) = \frac{1}{C\frac{2}{T}\frac{z-1}{z+1}}\left(I(z) - \frac{V(z)}{R}\right) \tag{5.43}$$

Rearranging produces

$$2C(z-1)V(z) = T(z+1)\left(I(z) - \frac{V(z)}{R}\right) \tag{5.44}$$

As before, we convert the above equation in the advance operator z to an equation in the delay operator z^{-1}:

$$2C(1 - z^{-1})V(z) = T(1 + z^{-1})\left(I(z) - \frac{V(z)}{R}\right) \tag{5.45}$$

The transformed difference equation will be

$$2Cv[n] - 2Cv[n-1] = Ti[n] + Ti[n-1] - \frac{T}{R}v[n] - \frac{T}{R}v[n-1] \tag{5.46}$$

The present sample of the output of the digital model can be expressed in terms of the present sample of the input and the past samples of the input and the output:

$$v[n] = \frac{Ti[n] + Ti[n-1] + 2Cv[n-1] - \frac{T}{R}v[n-1]}{2C + \frac{T}{R}} \tag{5.47}$$

This is our new model to simulate iteratively for $n = 0, 1, 2, \ldots$.

The change in the code for the lossy capacitor is minimal. The implementation of the digital model changes and there is another additional variable R which is the parallel loss resistor.

```python
import numpy as np
import matplotlib.pyplot as plt

C = 10.0e-6             # Farad
R = 2000.0             # Ohm
Imag = 5.0               # Amps
f = 50                   # Hertz
omega = 2*np.pi*f        # rad/s
Tg = 1.0e-6            # microseconds
t = np.arange(0, 1.0, Tg)
i = Imag*np.sin(omega*t)
```

```
T = 200.0e−6    # Sampling  time  period  of  200  mico
  secs  or  5  kHz
no_of_samples  =  int (T/Tg)
t_samples  =  t [:: no_of_samples ]
i_samples  =  i [:: no_of_samples ]

v_samples  =  np . zeros ( i_samples . size )

u  =  np . zeros ( 2 )
y  =  np . zeros ( 2 )

for  i_index ,  i_val  in  np . ndenumerate ( i_samples ):
    u[0]  =  i_val
    y[0]  =  ( T*u[0]  +  T*u[1]  +  2*C*y[1]  −  T*y[1]/R )  /
    (2*C  +  T/R)
    u[1]  =  u[0]
    y[1]  =  y[0]
    v_samples [ i_index ]  =  y[0]

plt . plot ( t ,  i ,  label = ' orig ' ,  ds = ' steps ' )
plt . plot ( t_samples ,  i_samples ,  label = ' samp ' ,  ds = ' steps ' )
plt . xlim ([0 ,  0.02])
plt . title ( ' Input  current ' )
plt . xlabel ( ' Time ' )
plt . ylabel ( ' Amps ' )
plt . legend ()

plt . figure ()
plt . plot ( t_samples ,  i_samples ,  label = ' input ' ,
  ds = ' steps ' )
plt . plot ( t_samples ,  v_samples ,  label = ' output ' ,
  ds = ' steps ' )
plt . xlim ([0 ,  0.08])
plt . title ( ' Input  and  output ' )
plt . xlabel ( ' Time ' )
plt . ylabel ( ' Amps  and  Volts ' )
plt . legend ()

plt . show ()
```

Figure 5.14 shows the simulation results when the loss resistor is added to the model. The current has the same initial dc offset but this can be seen to decay to zero in a few cycles. It should be noted that the value of $R = 2000$ Ohm chosen is quite small for a loss resistance and such a resistance will result in a very large unrealistic loss. A more realistic value would be $R = 50,000$ Ohm. This result is

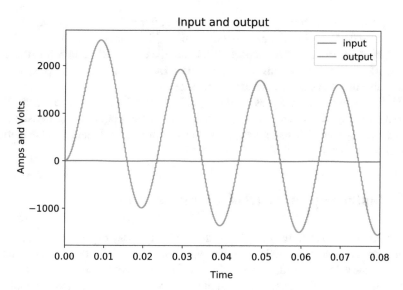

Fig. 5.14 Dc offset decaying with loss dissipation

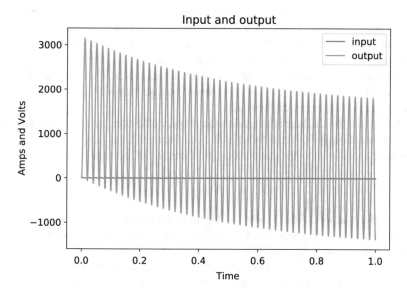

Fig. 5.15 Dc offset decaying gradually

shown in Fig. 5.15 where the dc offset in the current is seen to decay gradually and even after 1 s, has not fully decayed.

Next question, how long will the dc offset take to die away? This is determined by the time constant of the capacitor–resistor circuit. For a circuit with only a capacitor and a resistor across it, the time constant is RC—the product of their parametric values. The time needed for the circuit to settle after a transient (and start-up is a transient) is 4RC. A circuit is said to have settled if the output of the circuit is within a margin of 2% from the final expected value. In the case of R = 2000 Ohm, the time constant is 0.02 s and the settling time is 0.08 s. In the case of R = 50 000 Ohm, the time constant is 0.5 s and the settling time is 2 s.

In this section, we have presented a digital model for a practical capacitor. This behaves exactly like a real physical capacitor in a circuit both during transients as well as steady state. We can now repeat the same process to obtain a mathematical model for a lossy inductor in the next section.

5.9   Mathematical Model of a Lossy Inductor

In the previous section, we modeled and simulated a lossy capacitor. We saw how connecting a resistor in parallel with the capacitor captured the losses in the electric field and these losses resulted in the decay of the dc offset in the output current. In this section, we need to generate a digital model for a lossy inductor by including a loss resistor in the inductor as well [3].

In the case of an inductor, the loss is modeled a bit differently with respect to the capacitor. In the case of the capacitor, the loss is related to the electric field. And the electric field in the di-electric is directly proportional to the voltage across the capacitor terminals. Therefore, to model the loss of the capacitor, the sensible thing to do was to connect a resistance across the capacitor. This resistor will draw a current as long as there is a voltage across the capacitor which in turn means there is an electric field across the terminals.

In an inductor, the loss is related to the magnetic field in the core. The magnetic field is linked to the current through the inductor with the flux in the core being directly proportional to the current flowing through the inductor.

$$\phi \propto i \tag{5.48}$$

Therefore, to model the loss in the inductor, we connect a resistor in series with it. As long as there is a current through the inductor, there will be a magnetic field, and this current will also produce a loss in the series resistance. A very analogous concept to the capacitor except that the resistance is in series with the inductor. But that has to do with the form of energy that is being stored in the inductor as an element as compared to the capacitor. This is shown in Fig. 5.16.

Fig. 5.16 Lossy inductor with resistor in series

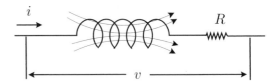

The equation for the inductor will change as the drop across the series resistor will need to excluded to apply the integral to the voltage drop across the inductor alone:

$$i = \frac{1}{L}\int (v - iR)dt \tag{5.49}$$

We can perform Laplace Transform on the above equation to convert it to the continuous frequency domain:

$$I(s) = \frac{1}{sL}[V(s) - I(s)R] \tag{5.50}$$

Converting from continuous frequency to digital frequency domain using Bilinear Transformation:

$$I(z) = \frac{1}{L\frac{2}{T}\frac{z-1}{z+1}}[V(z) - I(z)R] \tag{5.51}$$

To obtain a time domain difference equation in terms of signal samples, a few steps of rearrangement and simplification:

$$2L(z - 1)I(z) = T(z + 1)(V(z) - I(z)R) \tag{5.52}$$

$$2L(1 - z^{-1})I(z) = T(1 + z^{-1})(V(z) - I(z)R) \tag{5.53}$$

The transformed difference equation will be

$$2Li[n] - 2Li[n - 1] = Tv[n] + TV[n - 1] - Ti[n]R - Ti[n - 1]R \tag{5.54}$$

The final expression for the present sample of the output current with respect to the present sample of the input voltage and the past samples of input voltage and output current is

$$i[n] = \frac{Tv[n] + Tv[n - 1] + 2Li[n - 1] - Ti[n - 1]R}{2L + TR} \tag{5.55}$$

The code for the lossy inductor needs minor modifications from the code of a pure inductor. An additional resistor variable R is assigned. The final implementation of the inductor changes within the loop.

```python
import numpy as np
import matplotlib.pyplot as plt

L = 1.0e-3                # Henry
R = 0.05                  # Ohm
```

```python
Vmag = np.sqrt(2)*240      # Volts
f = 50                              # Hertz
omega = 2*np.pi*f              # rad/s
Tg = 1.0e-6                     # microseconds
t = np.arange(0, 1.0, Tg)
v = Vmag*np.sin(omega*t)

T = 200.0e-6    # Sampling time period of 200 mico secs
   or 5 kHz
no_of_samples = int(T/Tg)
t_samples = t[::no_of_samples]
v_samples = v[::no_of_samples]

i_samples = np.zeros(v_samples.size)

u = np.zeros(2)
y = np.zeros(2)

for v_index, v_val in np.ndenumerate(v_samples):
    u[0] = v_val
    y[0] = ( T*u[0] + T*u[1] + 2*L*y[1] - T*y[1]*R ) /
    (2*L + T*R)
    u[1] = u[0]
    y[1] = y[0]
    i_samples[v_index] = y[0]

plt.plot(t, v, label='orig', ds='steps')
plt.plot(t_samples, v_samples, label='samp', ds='steps')
plt.xlim([0, 0.02])
plt.title('Input voltage')
plt.xlabel('Time')
plt.ylabel('Volts')
plt.legend()

plt.figure()
plt.plot(t_samples, v_samples, label='input',
ds='steps')
plt.plot(t_samples, i_samples, label='output',
ds='steps')
plt.xlim([0, 0.08])
plt.title('Input and output')
plt.xlabel('Time')
plt.ylabel('Volts and Amps')
plt.legend()
```

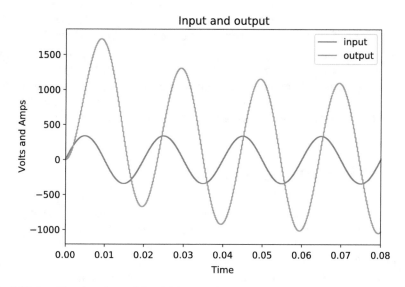

Fig. 5.17 Dc offset decaying rapidly with large series resistance

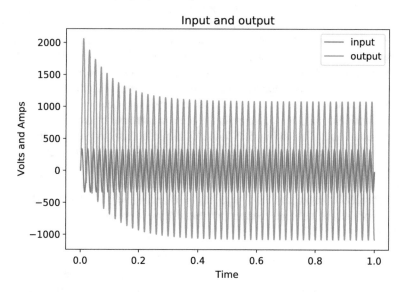

Fig. 5.18 Dc offset decaying gradually

plt.show()

The results are shown in Figs. 5.17 and 5.18. Figure 5.17 shows the rapid decay of the dc offset in the output current with a series resistance R = 0.05 Ohm. Such a large resistance is usually not realistic. Figure 5.17 shows the gradual decay of the dc offset in the output current with a smaller series resistance R = 0.01 Ohm. The time constant of an inductor–resistor (L-R) circuit is $\frac{L}{R}$. The settling time is four times the time constant i.e. $4\frac{L}{R}$. In the first case with R = 0.05 Ohm, the settling time was 0.08 s. In the second case with R = 0.01 Ohm, the settling time is 0.4 s.

In this section, we have created a digital model for a practical inductor that behaves exactly the way a real inductor does in a circuit. With this, we now have the basic building blocks to develop a digital model for a LC filter.

5.10   Modeling the LC Filter

In the past few sections, we developed the digital models for capacitors and inductors, as well as included loss components to generate models that behave like practical components. In this section, we will generate the digital model for the LC filter [3]. We could use the same approach as before of applying the Laplace Transform to the differential equations that describe the circuit. However, we will introduce another convenient approach that can be used to generating transfer functions for filters with several inductors and capacitors.

Figure 5.19 shows the LC filter which we have briefly discussed in the previous chapter. We have included a resistor R in series with the inductor. We could include a parallel resistor across the capacitor C as well to model the loss in both the inductor and the capacitor. However, to model a practical LC filter, we need to include a loss component which is represented by the resistor R and this resistor alone is sufficient to capture the loss in the filter. Additional resistors will only complicate the final equation and therefore, we will choose the minimal number of components.

The circuit on the left in Fig. 5.19 is the LC filter. In order to generate an equation for the output voltage v_o with respect to the input voltage v_{in}, we could write the differential equations and express the output voltage as a double integral of the input as we did in the previous chapter. However, if we assume a system at rest with all initial conditions to be zero i.e. $i = 0$, $v_{in} = 0$ and $v_o = 0$ at time $t = 0$, we can directly convert the elements to their frequency domain counterparts as shown in the right circuit of Fig. 5.19.

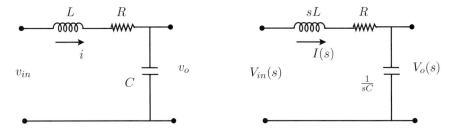

Fig. 5.19 LC filter with frequency domain representation

The circuit on the right of Fig. 5.19 can be solved just like a conventional circuit except that now even inductors and capacitors can be treated just like resistors. Therefore, the current in the frequency domain s is

$$I(s) = \frac{V_{in}(s)}{R + sL + \frac{1}{sC}} = \frac{sC}{LCs^2 + RCs + 1} V_{in}(s) \tag{5.56}$$

The output voltage can be expressed as

$$Vo(s) = \frac{I(s)}{sC} = \frac{V_{in}(s)}{LCs^2 + RCs + 1} = \frac{1}{LC} \frac{1}{s^2 + \frac{R}{L}s + \frac{1}{LC}} V_{in}(s) \tag{5.57}$$

We can convert this to the discrete domain using Bilinear Transformation (5.4):

$$Vo(z) = \frac{1}{LC} \frac{1}{\left(\frac{2}{T}\frac{z-1}{z+1}\right)^2 + \frac{R}{L}\left(\frac{2}{T}\frac{z-1}{z+1}\right) + \frac{1}{LC}} V_{in}(z) \tag{5.58}$$

As before, this needs expansion and rearrangement. The end result is

$$V_o(z) = \frac{\frac{1}{LC}(z^2 - 2z + 1)}{\left[\left(\frac{2}{T}\right)^2 + \frac{2}{T}\frac{R}{L} + \frac{1}{LC}\right]z^2 + \left[-2\left(\frac{2}{T}\right)^2 + \frac{2}{LC}\right]z + \left[\left(\frac{2}{T}\right)^2 - \frac{2}{T}\frac{R}{L} + \frac{1}{LC}\right]} V_{in}(z) \tag{5.59}$$

The above equation is a bit messy primarily due to the second order term in s. Naturally, higher order terms in s such as s^3, s^4, ... will be worse to solve and rearrange. In the next chapter, we will avoid these substitutions and introduce a Python command that automatically converts transfer functions from the continuous s domain to the digital z domain.

It is a simple matter to divide both numerator and the denominator by z^2 to express the above equation in terms of the delay operators z^{-1} and z^{-2} instead of the advance operators z^1 and z^2. The final time domain expression in terms of signal samples is

$$\left[\left(\frac{2}{T}\right)^2 + \frac{2}{T}\frac{R}{L} + \frac{1}{LC}\right]v_o[n] + \left[-2\left(\frac{2}{T}\right)^2 + \frac{2}{LC}\right] + v_o[n-1] \tag{5.60}$$

$$+ \left[\left(\frac{2}{T}\right)^2 - \frac{2}{T}\frac{R}{L} + \frac{1}{LC}\right]v_o[n-2] = \frac{1}{LC}(v_{in}[n] - 2v_{in}[n-1] + v_{in}[n-2]) \tag{5.61}$$

The above digital model can be programmed in exactly the same way as with inductors and capacitors. We will have a large list of variable declarations for the inductor L, resistor R and capacitor C. Besides this, another significant change is that we now have two past samples for both input $v_{in}[n-1]$, $v_{in}[n-2]$ and output

$v_o[n-1]$, $v_o[n-2]$. Therefore, the input and output variables will be declared as arrays of three elements instead of just two—0th element as the present sample, 1st element as the first past sample $n-1$ and 2nd element as the second past sample $n-2$.

```python
import numpy as np
import matplotlib.pyplot as plt

t_duration = 1.0
t_step = 1.0e-6

no_of_data = int(t_duration/t_step)

# Time array which is integers from 0 to 1 million -1
time_array = np.arange(no_of_data)*t_step

C = 1000.0e-6             # 1000 microFard capacitor
R = 0.05                  # 0.05 Ohm resistance as loss
L = 0.001                 # 1 milli Henry inductor
frequency = 50.0          # Hertz
omega = 2*np.pi*frequency # rad/s
mag = np.sqrt(2)*240.0    # Voltage

ip_voltage_signal = mag*np.sin(time_array*omega)

t_sample = 200.0e-6       # 5kHz
no_of_skip = int(t_sample/t_step)

tsample_array = time_array[::no_of_skip]
ip_voltage_samples = ip_voltage_signal[::no_of_skip]

# Initialize our output
op_voltage_samples = np.zeros(ip_voltage_samples.size)

# Filter input
u = np.zeros(3)
y = np.zeros(3)

# Calculate our output
for volt_index, volt_value in np.ndenumerate
(ip_voltage_samples):

    # Implementation of the filter
    u[0] = volt_value
    y[0] = ( (1/(L*C))*(u[0] + 2*u[1] + u[2]) - \\
```

```
(  −2*(2/t_sample)*(2/t_sample) + (2/(L*C)) )*y[1] − \\
(  (2/t_sample)*(2/t_sample) − (R/L)*(2/t_sample) + \\
(1/(L*C)) )*y[2] ) / ( (2/t_sample)*(2/t_sample) + \\
(R/L)*(2/t_sample) + (1/(L*C)))

    u[2]  =  u[1]       # u(n−2) = u(n−1)
    y[2]  =  y[1]       # y(n−2) = y(n−1)
    u[1]  =  u[0]       # u(n−1) = u(n)
    y[1]  =  y[0]       # y(n−1) = y(n)

    op_voltage_samples[volt_index] = y[0]
    # End of filter implementation

plt.plot(time_array, ip_voltage_signal, \
         label='full', ds='steps')
plt.plot(tsample_array, ip_voltage_samples, \
         label='sample', ds='steps')
plt.title('Input voltage with sampling effect')
plt.xlabel('Time')
plt.ylabel('Volts')
plt.legend()

plt.figure()
plt.plot(tsample_array, op_voltage_samples, \
         label='output', ds='steps')
plt.title('Output voltage')
plt.xlabel('Time')
plt.ylabel('Volts')
plt.legend()

plt.figure()
plt.plot(tsample_array, ip_voltage_samples, \
          label='input', ds='steps')
plt.plot(tsample_array, op_voltage_samples, \
          label='output', ds='steps')
plt.title('Input and output')
plt.xlabel('Time')
plt.ylabel('Volts')
plt.legend()

plt.show()
```

In this section, we presented the digital model for the LC filter along with the code. The simulation results of the LC filter will be presented in the next section with a detailed analysis of the performance of the filter.

5.11   Analysing the Performance of the LC Filter

In the previous section, we had developed the digital model for the LC filter and written the Python code. We had deferred the simulation results, as there are several aspects to the simulation we would like to examine. In this section, we will examine the performance of the digital LC filter using several test conditions.

To begin with, let us plot the results of executing the code in the previous section. Specifically, we have an input voltage of 240 V, 50 Hz (RMS), an inductor $L = 1$ mH, a capacitor $C = 1000\,\mu$F and a parasitic series resistor $R = 0.05$ Ohm. Figures 5.20, 5.21, and 5.22 are the simulation results. Figure 5.20 shows the output has an initial oscillation which dies down after a few cycles. Figure 5.21 shows a zoomed in plot of the input and output for the first few cycles. This plot shows that the output voltage appears to be distorted. Figure 5.22 shows the zoomed in plot of the input and output for the next few cycles where the output begins to look like a much smoother sine wave.

A few quick observations—What are the initial oscillations in the outer envelope and also the distortions in the output during the first few cycles? An LC circuit is a resonant circuit as will be discussed in detail in the next chapter. A resonant circuit will exhibit oscillations following a transient—and in this case, applying the initial input is a transient. The oscillations will die down if there is a loss component which in our case is modeled by the series resistance. Therefore, the oscillations which appear for the first few cycles seem to disappear.

Next observation—How is the output of the filter greater than the input? An oscillator such as an LC filter can have an amplification action on signals of

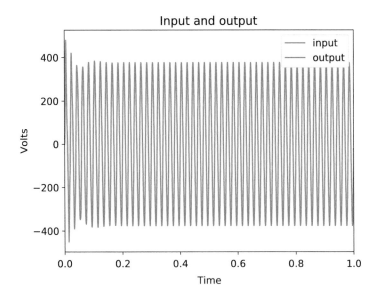

Fig. 5.20 Input and output envelopes of the LC filter

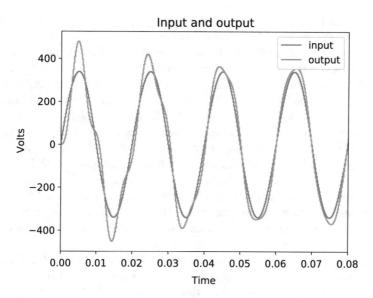

Fig. 5.21   Zoomed in input and output for first few cycles

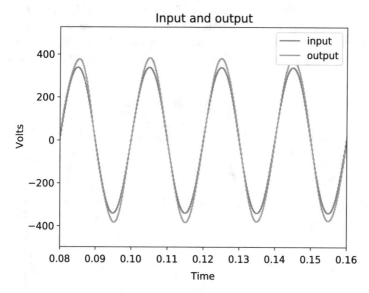

Fig. 5.22   Zoomed in input and output in steady state

certain frequencies. In the next chapter, this will be discussed in detail using frequency response analysis. However, for now, the amplification that we are seeing particularly in the plot Fig. 5.22 is not a bug. An example of a similar phenomenon is the Ferranti effect where a long transmission line that is lightly loaded has a higher voltage on the receiving end than on the sending end due to the effect of charging capacitors in the model of the transmission line.

With these few observations, the next step would be to try and examine the filtering effect of the model. In order to do so, we need to modify the input to the model. In the above case, we had a sine wave of 50 Hz as the input. To this waveform, let us add a 10% 650 Hz sine wave or 13th harmonic. Figures 5.23 and 5.24 show the input and the output for the first few cycles of transient and the next few cycles as steady state is achieved. In both plots, the input voltage can be seen to be fairly noisy as it is now a combination of a 50 Hz signal and a 650 Hz signal. The output however is still a sine wave that appears (particularly in Fig. 5.24) to be smooth and devoid of the higher frequency component. This is the filtering performance of the LC filter and the digital model that represents this LC filter. The filter allows low frequency components (50 Hz) in the input to pass through to the output but prevents higher frequency components (650 Hz) from appearing in the output.

This performance of the LC filter shown in Figs. 5.23 and 5.24 is incredibly useful when a particular signal contains noise but we wish to extract a signal free of the noise before processing it. Now that we have seen the LC filter in action, the next question would be—what frequencies pass through and what frequencies do not? Let us consider another input signal. This time, to the 50 Hz signal, let us add a 250 Hz (5th) harmonic instead. Figures 5.25 and 5.26 show the simulation results in this case. From both of the plots, it is immediately clear that the output is not smooth as it was with the previous case. Now the output signal appears to have some of the

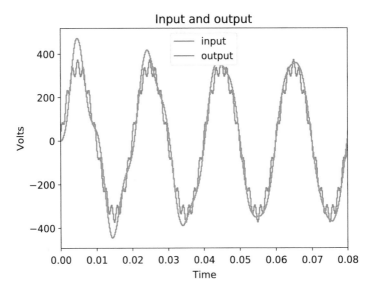

Fig. 5.23 Input and output with harmonics—first few cycles

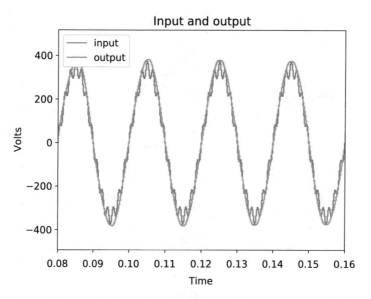

Fig. 5.24 Input and output with harmonics – in steady state

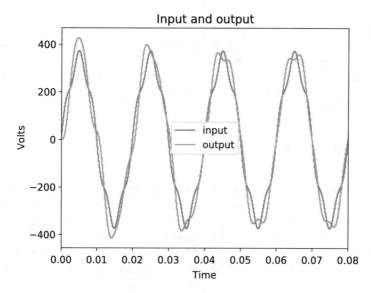

Fig. 5.25 Input and output with harmonics—first few cycles

250 Hz component superimposed to it. This is a case of the filtering action not being sufficient since the output is not the smooth 50 Hz sine wave that we got before.

With these simulation results, we have examined the basic working of the digital model of a LC filter. We have seen where the LC filter is effective and where it is not effective. However, these simulation results have opened a number of questions.

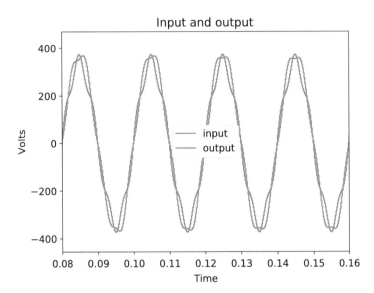

Fig. 5.26 Input and output with harmonics—in steady state

How do we ensure that a particular frequency component appears or does not appear in the output? Can we control the magnitude of the output to ensure that certain frequencies are preserved as they are in the input without any amplification or attenuation? The answers to all these questions will be provided in the next chapter where a detailed process for designing filters will be provided.

5.12   Conclusions

In this chapter, we finally began with some code implementations. The objective of this chapter was to ease the reader into the process of how to implement digital filters. The chapter chose the basic components—inductors and capacitor, to describe how we can begin with the well-known time differential equations, use Laplace Transform and subsequently convert the transformed equations to the digital domain. We use our knowledge from the previous chapter about how we can perceive frequencies in transformed equations as operators allowing us to generate time domain difference equations in terms of signal samples.

We built mathematical models step by step and combined them with the implementations using Python code. We used simulation results to analyse the entire process—sampling of signals, calculation of the output at every sample and the potential translation of the process to a practical hardware implementation. We concluded from the simulation results that our digital models were accurate and reflected how the analog components would behave. We finally synthesized and implemented the digital model for the LC filter and through simulation results examined the concept of filtering.

The purpose of this chapter was to perform a deep dive into the process of digital filtering using the very simple examples of capacitors, inductors and LC filters. Digital filtering can achieve far more complex objectives as will be described in the next few chapters. However, directly jumping to those concepts without tearing apart the process for simple analog circuits that we electrical engineers are comfortable with, causes DSP to become a confusing topic. In the next few chapters, we will gradually ease away from analog circuits and examine how a pure digital approach can provide incredibly powerful tools to design advanced filters.

References

1. Mitra, S. K. (1998). *Digital Signal Processing: A Computer-Based Approach*. McGraw-Hill.
2. Moschytz, G. S. (2019). *Analog Circuit Theory and Filter Design in the Digital World* (1st ed.). Springer International Publishing.
3. Ramo, S., Whinnery, J. R., & Duzer, T. V. (1994). *Fields and Waves in Communication Electronics* (3rd ed.). Wiley.

Chapter 6
Frequency Response Analysis

6.1 Introduction

In the previous chapter, we had derived the transfer functions for basic circuit components such as the inductor and capacitor so as to arrive at the transfer function of the LC filter which is a fairly popular analog filter. We had used our theoretical background from the earlier chapters to convert the continuous time transfer function to the digital domain so as to implement the LC filter transfer function as a difference equation. Using a simulation, we had examined the operation of the digital LC filter implementation. Specifically, we had introduced signals with different frequencies in the input signal to notice how the output signal changed.

The simulation raised a few very interesting questions. We saw how some frequency components in the input were almost eliminated in the output. However, some other frequency components appeared in the output even if they had been modified by the digital filter. With the simulation results, it became fairly obvious that a filter such as the LC filter behaves differently for signals of different frequency. In general, every filter is expected to behave differently for different frequencies. This frequency dependent behaviour is not random but is very well defined and can be studied in great detail. Studying how a transfer function behaves for different frequencies comprises studying the frequency response characteristics of the transfer function.

In this chapter, we will introduce Bode plots which are the most popular and probably the most powerful tool to study frequency response characteristics. With a basic introduction to the theory of Bode plots, this chapter will examine the Python function that can be used to generate Bode plots. We will then use these Python commands to study the frequency response characteristics of the inductor and the LC filter. Using the frequency response plots, we will interpret the behaviour of the LC filter that we observed in the previous chapter. We will then introduce the

© The Editor(s) (if applicable) and The Author(s), under exclusive license to
Springer Nature Switzerland AG 2020
S. V. Iyer, *Digital Filter Design using Python for Power Engineering Applications*,
https://doi.org/10.1007/978-3-030-61860-5_6

concept of generalized transfer functions and how these can be used as building blocks to design digital filters without any reference to analog circuits.

6.2   A New Look at Complex Numbers

In order to be able to understand frequency response analysis, we need to look at an alternate manner of expressing complex numbers [1]. Let us take an example of a complex number:

$$a = 2 + j3 \tag{6.1}$$

When we were looking into Euler's equation, we saw that a complex number can be represented on a plane with the x-axis being the real axis and the y-axis being the imaginary axis. We can do that here as well and for any complex number expressed in the $Re + jIm$ form with its real and imaginary part. A complex number can also be represented by another set of parameters which has more of a physical significance. The magnitude of the complex number is defined as

$$|a| = \sqrt{2^2 + 3^2} = \sqrt{Re\{a\}^2 + Im\{a\}^2} \tag{6.2}$$

The phase angle or the argument of a complex number can be written as

$$\arg(a) = \tan^{-1}\left(\frac{3}{2}\right) = \tan^{-1}\left(\frac{Im\{a\}}{Re\{a\}}\right) \tag{6.3}$$

To understand this, take a look at the way the complex number is represented in the two dimensional plane as shown in Fig. 6.1. As before, the real part of the complex number is along the real axis and the imaginary part is along the imaginary axis. This results in a phasor which has a magnitude equal to the magnitude of the complex number and a phase angle which is the angle of the phasor with the positive real axis. It is important to note that the phase angle is always defined as the angle with respect to the positive real axis and measured in the counter-clockwise sense.

$$x = a + jb \tag{6.4}$$

$$|x| = \sqrt{a^2 + b^2} \tag{6.5}$$

$$\arg(x) = \tan^{-1}\left(\frac{b}{a}\right) \tag{6.6}$$

With this we are now looking at complex numbers as phasors—with a magnitude and a phase angle.

Fig. 6.1 Expressing a complex number as a phasor

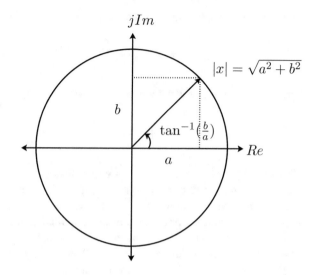

6.3 Transfer Functions: Magnitude and Phase Angle

In the previous section, we represented complex numbers as phasors with a magnitude and a phase angle. In this section, we will use this to represent transfer functions as phasors by first deriving the magnitude of any transfer function [2]. As an example, let us look at the current through the lossy inductor in the frequency domain:

$$I(s) = \frac{V(s)}{sL + R} \tag{6.7}$$

As discussed before in the previous chapter, s is the complex frequency:

$$s = \sigma + j\omega \tag{6.8}$$

In the previous chapter, we have also examined the physical significance of the real and imaginary parts of the frequency s when used to express a sinusoid. The real part determines the stability of the sinusoid while the imaginary part determines the frequency of the sinusoid. In signal processing, we usually omit the real part σ of the frequency. The reason is that signal processing is to deal with how a system behaves as the frequency changes while stability analysis is dealt with separately. So, in signal processing analysis, it is usual to approximate s as

$$s = j\omega \tag{6.9}$$

If we substitute this in the equation of current of (6.7):

$$I(j\omega) = \frac{V(j\omega)}{j\omega L + R} \tag{6.10}$$

The magnitude of this current expression will be

$$|I(j\omega)| = \frac{|V(j\omega)|}{\sqrt{(\omega L)^2 + R^2}} \tag{6.11}$$

Theoretically, the magnitude of an expression such as the one above should be expressed after simplifying it to a $Re + jIm$ form. However, the magnitude of a complex number that is a product or a division of other complex numbers can be expressed as a product or division of the magnitudes of each complex number.

Phase angles, on the other hand, are a bit more complex due to the inverse tangent that appears. There is an approximation that can be used for calculating the phase angle. To be able to understand how this approximation is arrived at, let us consider this simple example:

$$x = \frac{a + jb}{c + jd} \tag{6.12}$$

If we multiply both numerator and denominator by the complex conjugate of the denominator $c - jd$:

$$x = \frac{(a + jb)(c - jd)}{c^2 - j^2 d^2} \tag{6.13}$$

From the definition of complex numbers, we know that

$$j = \sqrt{-1} \tag{6.14}$$

So,

$$j^2 = -1 \tag{6.15}$$

Using this in the modified expression:

$$x = \frac{(a + jb)(c - jd)}{c^2 + d^2} \tag{6.16}$$

We have removed a complex number from the denominator. This technique of multiplying both numerator and the denominator by the complex conjugate of the denominator is a very commonly used method to move all complex numbers to the numerator. The numerator also can be expanded:

$$x = \frac{ac + bd + j(bc - ad)}{c^2 + d^2} \tag{6.17}$$

Following this rearrangement, the phase angle can be calculated as

$$\arg(x) = \tan^{-1} \frac{bc - ad}{ac + bd} \tag{6.18}$$

It is however not very convenient to perform these computations for every transfer function. We need a way to compute phase angle of a transfer function with respect to the phase angle of every individual component the way we did for the magnitude. For this purpose, we need to examine the nature of the tangent function. The tangent of an angle increases from 0 for an angle of 0 radian to ∞ for an angle of $\frac{\pi}{2}$. The function of course is non-linear. However, the tangent of angles that are small is approximately equal to the angle itself. As an example:

$$\tan 30° = \tan \frac{\pi}{6} = 0.5773 \tag{6.19}$$

and since:

$$\frac{\pi}{6} = 0.5235 \tag{6.20}$$

This approximation is not too bad. In general for angles less than 45° or $\frac{\pi}{4}$ radians, this approximation can be applied.

The above approximation can also be applied for the inverse tangent that we use for calculating phase angles.

$$\tan^{-1} x \tag{6.21}$$

For small values of x, the inverse tangent will result in an angle that is approximately equal to the value x itself. For large values of x, the inverse tangent will tend towards 90° in a manner called asymptotic. How do we use this knowledge to simplify the expression for phase angle?

In the expression for the phase angle:

$$\arg(x) = \tan^{-1} \frac{bc - ad}{ac + bd} \tag{6.22}$$

The result will be approximately

$$\arg(x) \approx \frac{bc - ad}{ac + bd} \tag{6.23}$$

as long as the right hand side is less than $\frac{\pi}{4}$. This is a huge simplification as we are able to express the phase angle without the inverse tangent function. However, we need to go one step further and express the phase angle with respect to the phase angles of the individual complex numbers.

The phase angles of the complex numbers that form the expression can also use the approximation:

$$\arg(a + jb) \approx \frac{b}{a} \tag{6.24}$$

$$\arg(c - jd) \approx -\frac{d}{c} \tag{6.25}$$

If we take the following expression:

$$\frac{b}{a} - \frac{d}{c} = \frac{bc - ad}{ac} \tag{6.26}$$

This expression is very close to

$$\frac{bc - ad}{ac + bd} \tag{6.27}$$

All of the above playing around with equations may seem like guesswork, but quite surprisingly many advancements in science were the result of scientists just playing around with expressions. The difference between (6.27) and (6.26) is the term bd. If the phase angles of the individual complex numbers are small, this would imply that $\frac{b}{a}$ and $\frac{d}{c}$ are small. This implies

$$b << a \tag{6.28}$$

$$d << c \tag{6.29}$$

Therefore, the product also satisfies

$$bd << ac \tag{6.30}$$

Resulting in the approximation:

$$\frac{bc - ad}{ac + bd} \approx \frac{bc - ad}{ac} \tag{6.31}$$

The result of all these approximations is that the phase angle of the complex number:

$$\arg\left(\frac{a + jb}{c + jd}\right) \approx \arg(a + jb) - \arg(c + jd) \tag{6.32}$$

The condition here is—if the individual phase angles are small. In general, the product and divisions of complex numbers will appear as summation and difference of phase angles of the individual components.

$$\arg\left(\frac{(a_1 + jb_1)(a_2 + jb_2)}{(c_1 + jd_1)(c_2 + jd_2)}\right) \approx \arg(a_1 + jb_1) + \arg(a_2 + jb_2)$$

$$-\arg(c_1 + jd_1) - \arg(c_2 + jd_2) \qquad (6.33)$$

With this background, the phase angle of the current through an inductor can be expressed as

$$\arg(I(j\omega)) \approx \arg(V(j\omega)) - \arg(j\omega L + R) \qquad (6.34)$$

With this background, we now can talk about the magnitude and phase angles of transfer functions in the next section where we will bring in the concept of frequency response.

This approximation makes it very convenient to express phase angles of complex transfer functions that may have many smaller transfer functions as will be shown soon. Manual calculation of phase angles is no longer necessary as we now have very convenient computational tools which will also be shown very soon. A knowledge of how phase angles can be approximated and how the plots of phase angles appear linear in later sections and also in the next chapter will make the entire process of frequency response analysis much more understandable.

6.4 Transfer Function: Frequency Response

In the previous section, we defined how any complex number can be represented as a phasor having a magnitude and a phase angle. We extended this definition to a complex number that would arise by performing the Laplace Transform on a time varying function of a physical quantity in a circuit such as the current through an inductor. With this background, in this section, we will introduce the concept of frequency analysis.

Let us go back to the inductor current expression from the previous section:

$$I(j\omega) = \frac{V(j\omega)}{j\omega L + R} \qquad (6.35)$$

It is important to note that signal processing requires the analysis of circuits (or mere components such as the inductor above). For the purpose of analysis, we write equations typically starting with time domain equations and then convert the equations using Laplace Transform to the frequency domain. A frequency domain equation such as the one above is ideal for signal processing, as in signal processing, we examine the behaviour of a system for different frequencies. The first step to system analysis is to decide the inputs to the system and the outputs from the system.

When talking about a system, in signal processing, we are talking about an equation in the frequency domain. For a circuit or a component, we could write a number of different equations and express it in a number of ways. Therefore, the system is like a box that we could flip over to a side or turn upside down. Depending on how we write equations for a system, the system will have inputs that are variables in the equations and will have outputs that are also variables in the equations. As an example, in the inductor equation above, the current $I(j\omega)$ could be the output and the voltage $V(j\omega)$ could be the input. In such a case, we could express the output with respect to the input as follows:

$$I(j\omega) = H(j\omega)V(j\omega) \tag{6.36}$$

where

$$H(j\omega) = \frac{1}{j\omega L + R} \tag{6.37}$$

is the transfer function of the system—the inductor. Again, this is just perspective. So, we could also have the voltage as the output and the current as the input, so as to write the equations as

$$V(j\omega) = H'(j\omega)I(j\omega) \tag{6.38}$$

where, the alternative transfer function $H'(j\omega)$ will be

$$H'(j\omega) = j\omega L + R \tag{6.39}$$

The transfer function will eventually be a complex number as is above. Therefore, a transfer function will have a magnitude and a phase angle. Both magnitude and phase angle are dependent on the frequency ω. In signal processing, the main question asked is how does a system behave as the frequency of inputs fed to it change? In layman's language, this is the equivalent of playing with the tuning knob of a radio and listening to channels. We can substitute different values of frequency ω into $H(j\omega)$ and calculate for each frequency, the magnitude and phase angle of the transfer function. This is frequency response analysis of a system.

Before getting into frequency response of the system in more detail, let us first examine why is the representation of a circuit or a system as a transfer function useful? Let us say, there is some frequency ω_1 which is of interest for some reason. Maybe, this frequency is found in the input and we would like to know how it appears in the output. We can calculate the magnitude and phase angle of $H(j\omega_1)$ as $|H(j\omega_1)|$ and $\arg(H(j\omega_1))$, respectively. How do we use this? We can express the magnitude of current with respect to the magnitude of the transfer function and the magnitude of the voltage at this frequency:

$$|I(j\omega_1)| = |V(j\omega_1)||H(j\omega_1)| \tag{6.40}$$

From the magnitude $|H(j\omega_1)|$, we could determine whether the magnitude of the current $|I(j\omega_1)|$ will be lower than or greater than the magnitude of the voltage $|V(j\omega_1)|$. And of course, we could precisely calculate the magnitude of current for a particular magnitude of voltage for the frequency ω_1 if for this frequency, the magnitude $|H(j\omega_1)|$ of the transfer function was known.

For the phase angle of the current, there is another relation that we have already described in the previous section:

$$\arg(I(j\omega_1)) = \arg(V(j\omega_1)H(j\omega_1)) = \arg(V(j\omega_1)) + \arg(H(j\omega_1)) \tag{6.41}$$

If, for the frequency ω_1, for a particular phase angle of the voltage $\arg(V(j\omega_1))$, if the phase angle of the transfer function $\arg(H(j\omega_1))$ is known, the phase angle of the current can be calculated by mere addition. Additionally, if the phase angle of the transfer function is greater than 0—$\arg(H(j\omega_1)) > 0$, the current will lead the voltage, and if $\arg(H(j\omega_1)) < 0$, the current will lag the voltage.

This would of course be obvious mathematically, but how does the concept of a transfer function help in these calculations if these calculations can be performed by just solving time varying functions as well? The advantage arises that transfer functions have magnitudes and phase angles that vary with frequency but these variations have been mapped out for most standard forms of transfer functions. Therefore, for any given system or circuit, the resultant transfer function will usually be an expression of a particular form. And this particular form has a magnitude and phase angle whose variation with frequency is well defined. Therefore, our calculations of the magnitude and phase angle of the output are now significantly simpler as the computation of the magnitude and phase angle of the transfer function is merely the result of well-established rules. This is what makes frequency response analysis so powerful as will be shown in the next few sections.

6.5 Bode Plots

In the previous section, we saw how the transfer function of a system is a function of frequency. We concluded the previous section with the statement that transfer functions tend to have standard forms which in turn result in well-defined behaviour of the systems with respect to changes in the magnitude and phase angle as frequency changes. In this section, we will introduce a very popular tool for studying the frequency response of a system.

A powerful tool in analysing how a system behaves with changing frequency is to plot the magnitude and phase angle of the system's transfer function with respect to frequency. At one glance, the behaviour of the system for a huge range of frequencies is now evident. The most famous frequency response plots are called Bode plots named after Hendrik Wade Bode [3]. A Bode plot of a transfer function consists of a plot of the magnitude and the phase angle with respect to frequency [2]. However, what makes the Bode plot different with respect to many other frequency response plots, is the representation of the frequency in the plots.

In signal processing, we usually want to see how a system behaves for a large range in frequency. This is because even if our frequency of interest may be only 50 or 60 Hz, you never know what frequencies could appear in a system. As an example, we might design a filter to remove 450 Hz components from a signal. But for some strange reason, this filter shows a very unstable and spiky response for frequencies close to 7000 Hz. It is possible, as every practicing engineering knows, that during implementation in the field, any frequency can appear. In that case, if this 7000 Hz frequency appears, the filter that we designed could result in oscillations in the system which is totally unacceptable. Remember, the laboratory is a controlled environment. The field is totally different and we must prepare for the unexpected.

What this implies is that even if the frequencies that we are interested in are fairly narrow, we must take into account a fairly wide range of frequencies just to be safe. To begin with, this poses an immediate problem. If our frequency range is from 0.1 Hz to 100 kHz, how are we going to plot this? Trying to plot this in a linear scale will be ridiculous as the resolution of the plot will be very poor making any kind of analysis very difficult and error prone. And for this reason, we come up with a logarithmic scale for frequencies.

In Bode plots, the x-axes of the magnitude and the phase angle plots is the logarithm of the frequency ω to the base 10:

$$x = \log_{10} \omega \tag{6.42}$$

What does this mean? When:

$$\omega = 1, \quad x = \log_{10} 1 = 0$$
$$\omega = 10, \quad x = \log_{10} 10 = 1$$
$$\omega = 100, \quad x = \log_{10} 100 = 2$$
$$\omega = 1000, \quad x = \log_{10} 1000 = 3$$
$$\omega = 100000, \quad x = \log_{10} 100000 = 5$$

There is a negative part of the x-axes as well:

$$\omega = 0.1, \quad x = \log_{10} 0.1 = -1$$
$$\omega = 0.01, \quad x = \log_{10} 0.01 = -2$$

A quick look at these numbers shows that the x-axis has been squeezed for the same frequency range. To represent frequencies from $\omega = 0.01$ to $\omega = 100000$, the x-axis is now from -2 to 5. This is way easier to represent on a graph. The next question—would this not hurt accuracy?

Well, yes it would. However, as will be seen very soon when we begin to generate Bode plots, drastic changes do not happen for small changes in frequency. A magnitude or a phase angle plot is not similar to a foreign currency exchange fluctuation where drastic changes can occur at regular intervals. In frequency response analysis, a vast majority of the changes in magnitude and phase angle occur over large changes in frequency. Therefore, the logarithmic scale produces frequency response plots that provide detailed insights about the transfer function of a system.

The next peculiarity about Bode plots is with respect to how the magnitude of a transfer function is defined. We have already seen with simple complex numbers, how the magnitude of a complex number can be calculated. For a transfer function $H(j\omega)$, the magnitude is defined as

$$|H(j\omega)| = \sqrt{Re(H(j\omega))^2 + Im(H(j\omega))^2} \tag{6.43}$$

This expression is arrived at using this expression:

$$|H(j\omega)| = \sqrt{H(j\omega)H^*(j\omega)} \tag{6.44}$$

where $H^*(j\omega)$ is the complex conjugate of $H(j\omega)$:

$$H^*(j\omega) = Re(H(j\omega)) - jIm(H(j\omega)) \tag{6.45}$$

In Bode plots, the magnitude of the transfer function is expressed in decibels. The use of decibels has its origin in telephony and as a unit of measurement of the power in an audio signal. However, the advantage of decibels is its flexibility in expressing magnitude or power in a signal as either a change or as an absolute. It is for this reason that decibels as a unit is very popular in signal processing and control as will be shown very soon. The magnitude of a transfer function in a Bode plot is expressed as

$$20\log_{10}|H(j\omega)| \tag{6.46}$$

At a first glance, this seems to be a needless complication just to conform with some historical definition of magnitude in sound technology which is around a 100 years old. Can't we move on? But, this definition does something very interesting to the magnitude of a transfer function that is comprised of smaller transfer functions. Suppose, we had

$$H(j\omega) = (a + jb\omega)(c + jd\omega) \tag{6.47}$$

The magnitude of this transfer function is

$$|H(j\omega)| = \sqrt{H(j\omega)H^*(j\omega)} = \sqrt{(a + jb\omega)(c + jd\omega)(a - jb\omega)(c - jd\omega)}$$
$$(6.48)$$

And thus, rearranging and separating the square roots, the magnitude of the transfer function is just the product of the magnitude of the individual components:

$$|H(j\omega)| = |a + jb\omega||c + jd\omega| \qquad (6.49)$$

This might seem simple enough, but when expressing the magnitude in decibels:

$$20\log_{10}|H(j\omega)| = 20\log_{10}|a + jb\omega||c + jd\omega| \qquad (6.50)$$

We find that the magnitude of the transfer function is a logarithm of products which we know is a sum of logarithms:

$$20\log_{10}|H(j\omega)| = 20\log_{10}|a + jb\omega| + 20\log_{10}|c + jd\omega| \qquad (6.51)$$

And similarly if the transfer function were to be a ratio of components:

$$H(j\omega) = \frac{a + jb\omega}{c + jd\omega} \qquad (6.52)$$

The magnitude in decibels will be a difference in logarithms:

$$20\log_{10}|H(j\omega)| = 20\log_{10}|a + jb\omega| - 20\log_{10}|c + jd\omega| \qquad (6.53)$$

By expressing the magnitude of the transfer function in decibels, if the transfer function was made up of smaller components and could be expressed as a product and ratio of these components, the magnitude in decibels will be the sum and difference of the magnitudes of the individual components in decibels. This is a huge simplification as quite often and as will be seen soon, many filters are a composition of transfer functions. The ability to be able to express the magnitudes as sums and differences results in a simplified analysis and easier predictions.

The phase angle of a transfer function is typically expressed in degrees and the calculation is the same as the calculation of the phase angle of any complex number. The true power of the Bode plots will only be evident once we begin to generate them for transfer functions. In the next chapter, we will use Bode plots exclusively as a tool for designing digital filters.

6.6   Getting Started with Scipy.Signal

In the previous sections, we had discussed about frequency response analysis and presented Bode plots as a tool to generate and analyse frequency response characteristics. Several years back, when we did not have the computation tools to generate Bode plots through programs, the theory behind Bode plots was used to generate them with pencil and paper. Fortunately, we now have packages in several software to generate Bode plots for complex transfer functions with just a few statements. In this section, we will discuss the tools available with SciPy with which we can generate Bode plots using Python code.

In this section, we will start exploring the signal package that comes with scipy. A handy reference to all the functions in this signal package can be found in the link:

https://docs.scipy.org/doc/scipy/reference/signal.html

The signal package is quite an interesting package and there are several functions available. I would encourage the reader to explore these functions and try them out in the examples we will undertake in this book. However, to build strong fundamentals, we will examine basic functions and focus on how they can be used to implement signal processing functions.

To begin with, let us represent a transfer function. Let us start with the transfer function of the inductor in the s-domain:

$$I(s) = V(s)H(s) \tag{6.54}$$

where:

$$H(s) = \frac{1}{R + sL} \tag{6.55}$$

To represent such a transfer function, we use the lti function in the signal package where LTI stands for Linear Time Invariant systems. To realize the above transfer function, we use this lti function as follows:

```
from scipy import signal
R = 0.1
L = 0.001
tf1 = signal.lti([1], [L, R])
```

Let us now dive into the details of this lti function. The lti function can be used in a number of different ways. In its simplest form, the lti function needs at least 2 arguments while it can have 3 or 4 arguments as well. In its simplest form, the lti function is used as follows:

```
tf = signal.lti(numerator, denominator)
```

Both numerator and denominator in the above command should be specified as lists or arrays (within [ ]). The numbers specified in the lists are the coefficients of s^k in descending order to s^0 or 1. This is best described through examples. For example:

```
[1, 1]
```

means $1 * s^1 + 1 * s^0 = s + 1$. If we change the coefficients:

```
[5, 3]
```

means $5 * s^1 + 3 * s^0 = 5s + 3$.

Higher order polynomials can be represented in this form as well. All that is needed to understand is that the element at index 0 in the list has the highest power while the element at the final index has the lowest power which is always $s^0 = 1$. Therefore:

[3 , 6 , 9 , 1]

means $3s^3 + 6s^2 + 9s^1 + s^0$. It is very important to note that the lti function will always search for the last element and take that as s^0. Therefore, the power of s decreases as we move from the element at index 0 to the final index. Moreover, the power of s will decrement by 1 for every consecutive element in the list. In case, there are some powers of s missing in a polynomial, these need to be taken into account in the corresponding list that represents the polynomial. So, suppose we have a function:

$$2s^7 + 3s^5 + 4s^3 + 2s^1 \tag{6.56}$$

Note how we have skipped all the even powers of s in this example. Any power of s could be absent in a polynomial. This would be translated to

[2 , 0 , 3 , 0 , 4 , 0 , 2 , 0]

We must specify 0 as the powers of s do not exist. lti will not know otherwise. All that lti knows is that the element at the last index is the s^0 power and this decreases until the element at index 0 which is the highest power. In this manner, we can specify polynomials in s for the numerator and denominator. An example:

$$\frac{s^3 + 4s^2 + 5}{2s^5 + 3s^4 + 5s^2 + 4s} \tag{6.57}$$

will be

tf2 = signal.lti ([1 , 4 , 0 , 5], [2 , 3 , 0 , 5 , 4 , 0])

In the above cases, the lti command has been used with two arguments which are numerator and denominator. However, the lti command can be used with two additional arguments as well:

tf = signal.lti (numerator , denominator , gain , dt)

The third argument "gain" is the gain pre-multiplier for the transfer function. The fourth argument is the sampling time in case we wish to define a transfer function in the digital domain. However, in this book, we will not use these two arguments. When a gain is required, we will modify the lists that represent the numerator and the denominator as this provides finer control over the transfer function representation when synthesizing higher order transfer functions. When a transfer function needs to be represented in the digital domain, we will first represent it in the continuous *s* frequency domain and then convert it to the digital *z* domain. The reader is encouraged to explore the lti command with these arguments as well.

With the above commands, we have described how we can represent a transfer function in frequency *s* using Python code with the signal package in SciPy. This was a basic introduction to get started. In the next section, we will begin to explore other properties of the lti function and how we can use this function to generate Bode plots.

6.7   Frequency Response of an Inductor

In the previous section, we saw how to represent a transfer function with the lti command in the signal package available with scipy. All we had done was described how to use the command and how we need to construct the arguments so as to represent a transfer function. However, the next question that arises is—how will this command help in frequency response analysis? It is important to understand that no one single command will perform every task. In this section, we will use this command to proceed with frequency analysis by using another command also available with the signal package – the bode command.

Before we start with the bode command, let us examine what the lti command produces in greater detail.

```
tf = signal.lti(numerator, denominator)
```

The output of the lti command is assigned to the variable tf. In Python, every such variable is an object. An object is a programming construct that became popular with the advent of Object Oriented Programming (OOP). In the older style of programming before OOP, data and code were usually separate entities. The code would use data, produce new data that would be pushed into the data pile to augment it further. In the OOP concept, data and code are not completely separate but rather combined together using a logical construct called a class. A class consists of data and code that operates the data. Several instance of a single class can be created and these instances are called objects. Each object can have data common to each other and also contain data unique to each and every object. Moreover, each object will have code that can operate on the data contained within the object. Let us examine this in more detail with the result of the lti command.

So, if we were to repeat the commands to create the inductor transfer function:

```
R = 0.1
L = 0.001
ind_tf = signal.lti([1], [L, R])
```

and ask the question, what is "ind_tf"? The answer would be

```
TransferFunctionContinuous(
array([1000.]),
array([  1., 100.]),
```

```
dt: None
)
```

And this of course, leads to the next question—what is this TransferFunctionContinuous? If we check the data type of the result "ind_tf":

```
type(ind_tf)
```

we get

```
scipy.signal.ltisys.TransferFunctionContinuous
```

TransferFunctionContinuous is a class inside the package ltisys inside the
package signal. As you can already see, the signal package contains several sub-
packages, functions and classes that are extremely useful in signal processing. The
lti command creates an object instance using the TransferFunctionContinuous class
with the arguments that we passed to the lti function. We passed a numerator and a
denominator as Python lists. The lti command standardizes this transfer function.
By standardization, we mean representing it in another form. However, the lti
command specifically rearranges the coefficients such that the highest power of s in
the denominator has the coefficient of 1. This is just a bit of mathematical jugglery
and should not be worrisome. As already stated, the lti function takes a gain and a
sampling time interval as third and fourth arguments. Since, we have not provided
that, TransferFunctionContinuous has "dt=None" just to signify that no sampling
time interval has been provided. But it is an optional argument just as gain is, which
is why TransferFunctionContinuous is successfully created. This is another concept
called overloading, where a class function which in this case is the constructor, can
be called with different arguments.

Now that we talked about how the lti command creates an object that is an
instance of the TransferFunctionContinuous class, let us examine the contents of
this class. As already stated, an object contains data and code that operates on the
data. To find out what are the data and methods in the object variable ind_tf above,
we use the following Python command:

```
dir(ind_tf)
```

This will list out many items. Let me list a few important ones here:

```
['bode', 'den', 'dt', 'freqresp', 'impulse',
 'inputs', 'num', 'output', 'outputs',
 'poles', 'step', 'to_discrete', 'to_ss',
 'to_tf', 'to_zpk', 'zeros' ]
```

In the above list, it is important to note that some are data which are called attributes
or properties and some are code which are called methods. As an example, if we
wish to know about the poles of this transfer function, we access it with the period
(.) delimiter:

```
ind_tf.poles
```

and the result will be

```
array([-100.])
```

The above result merely indicates that the transfer function has a single pole (an array with one element) of value -100. I would encourage the reader to verify this.

Before we start looking at other attributes and methods, let us pause to examine the usefulness of OOP. We had used the lti command with two arguments. The lti command returned an object which has been created (constructed) with these two arguments. And this object which was an instance of a class, now comes with a fairly vast collection of attributes and methods in addition to the data that we had specified. The numerator and denominator which we have passed to the lti function are available as the "num" and "den" attributes of ind_tf. So without writing any code, we can access several other aspects of the transfer function through the attributes and methods. This is the fundamental principle of OOP. When a particular type of data is used in a particular manner, defining a class with attributes and methods surrounding the core data allows for convenient information access. We get more information with less code.

With this little detour to talk about OOP and how the result of the lti command that we used in the previous section is related to OOP since it returns an object, we can get back to what we wanted to do—generate a Bode plot. To generate a Bode plot, we have another function in the signal package in scipy—the bode function. And this bode function merely takes as an argument the ind_tf object created by the lti function. In its simplest form, just use

```
signal.bode(ind_tf)
```

The bode function needs a system as an argument.

```
signal.bode(system)
```

In our case, we have been using a TransferFunctionContinuous object. However, the bode function takes other types of system representations as well.

Used with just one argument which is the system representation, the function will assume a range of frequencies ω for the frequency response characteristic in the Bode plot. Unless our intention is just to play around with Bode plot, we cannot just let the bode function assume a frequency range. Therefore, just like with the lti command, the bode command can be called with the frequency range as an optional argument. From the official documentation, the bode function can be used as follows:

```
signal.bode(system, w = None, n = 100)
```

The second argument is the frequency. There is also a third argument which is the number of frequency points to be chosen for the Bode plot if the frequency argument is not provided. In our cases, we will be specifying the frequency as a range object as all that the bode function needs is an iterable similar to an array. Therefore, an implementation as follows will be most convenient:

```
signal.bode(system, range(0.1, 10000))
```

In the above case, we are generating a Bode plot for a frequency range between 0.1 and 10000 rad/s. It is important to note that frequency specified is assumed to be in radian per second.

Now that we have discussed how to call the bode function, let us talk about how to generate a frequency response plot. The bode function will not create a plot as a diagram or figure. The bode function returns three arguments—a frequency array, the magnitude array in decibels and the phase angle array in degrees. All three arrays have the same length and are data points for the magnitude and phase angle plots that comprise the Bode plot. At first thought, why would we need the frequency array as an output of the bode function when we providing it as an argument? To begin with, the frequency is an optional argument, and therefore, even if the frequency is not provided to the function, the function will need to specify the frequency of each data point. Additionally, even if the frequency is specified as an argument, the output of the bode function will not be the same. The frequency that is generated as an output is the logarithm of the frequency since the x axis in the Bode plot is the logarithm of the frequency. With this said, the bode function is used as follows:

```
w, mag, phase = signal.bode(ind_tf, frequency_range)
```

where w, mag and phase are NumPy arrays.

Now that we have the x and y coordinates of every data point in the magnitude and phase angle plot, we can use Matplotlib to generate the visual Bode plot. In the previous chapter, we used the plt function in the pyplot package to plot waveforms. However, the Bode plot has the x axis as the logarithm of the frequency. The convenience of using the bode function is that the frequency produced as an output is the logarithm of the frequency of each data point. All that is needed is a plotting function that scales the x axis with a log scale. For this, there is a convenient function in the same pyplot package—the semilogx function:

```
import matplotlib.pyplot as plt
plt.semilogx(w, mag)
```

The above function will generate the magnitude plot. To generate the phase angle plot:

```
plt.semilogx(w, phase)
```

With the above discussion on how to generate a Bode plot consisting of the magnitude and phase angle responses, we can now describe the significance of these frequency response plots. In the next section, we will generate the Bode plots and analyse them.

6.8 Interpreting the Inductor Frequency Response

In the previous section, we examined the bode function that we will use to generate Bode plots. In this section, we will use the code from the previous section and interpret the frequency response plots that are produced. The objective of this analysis is to lay the foundation of how filters behave with changing frequency which will be the basis for filter design in the next chapter.

Let us list the entire code for generating the Bode plots:

```python
import numpy as np
import matplotlib.pyplot as plt
from scipy import signal
R = 0.1
L = 0.001
num_l = [1]
den_l = [L, R]
ind_tf = signal.lti(num_l, den_l)
w, mag, phase = signal.bode(ind_tf,
    np.arange(0.1, 100000))
plt.figure()
plt.title(''Magnitude'')
plt.xlabel(''Frequency'')
plt.ylabel(''dB'')
plt.grid()
plt.semilogx(w, mag)
plt.figure()
plt.title(''Phase angle'')
plt.xlabel(''Frequency'')
plt.ylabel(''Degrees'')
plt.grid()
plt.semilogx(w, phase)
plt.show()
```

The Bode plots that result from this code are shown in Figs. 6.2.

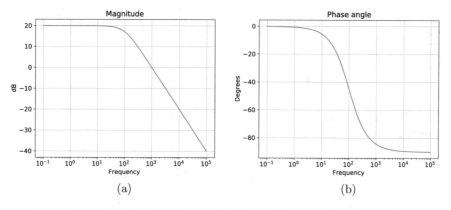

(a)   (b)

Fig. 6.2 Bode plots of the inductor transfer function. (**a**) Magnitude plot (**b**) Phase angle plot

The bode function is therefore a very convenient way to generate frequency response plots. Before we accept the result of the Bode plot as the final truth, let us try to double check the values that are on the plot. As always, we start with the inductor transfer function:

$$\frac{I(j\omega)}{V(j\omega)} = H(j\omega) = \frac{1}{0.001\,j\omega + 0.1} \tag{6.58}$$

There are many ways to adjust and rearrange the above equation. To examine the effect of frequency, the following rearrangement makes interpretation fairly simple as will be shown soon:

$$H(j\omega) = \frac{10}{0.01\,j\omega + 1} \tag{6.59}$$

The magnitude of the transfer function in decibels (dB) as used for the magnitude plot of the Bode plots is

$$20\log_{10} |H(jw)| = 20\log_{10} 10 - 20\log_{10} (|0.01\,j\omega + 1|) \tag{6.60}$$

We have merely used the property of the logarithm as a function—the logarithm of a ratio is the difference of the logarithms of the numerator and the denominator. But by doing so, we have separated the magnitude into two terms—the first term is independent of frequency while second term is dependent on frequency.

To fully understand the impact of frequency on the magnitude $|H(j\omega)|$ of the transfer function, let us examine separately this term while rearranging and expanding it:

$$20\log_{10} (|0.01\,j\omega + 1|) = 20\log_{10} \left(\left| j\frac{\omega}{100} + 1 \right| \right) = 20\log_{10} \left[\sqrt{\left(\frac{\omega}{100}\right)^2 + 1} \right] \tag{6.61}$$

For ω values much smaller than 100, the ratio $\frac{\omega}{100}$ is small (less than 1) and the square of that is even smaller—remember, the square of a number less than 1 is lesser than the number itself. So,

$$\left(\frac{\omega}{100}\right)^2 \ll 1 \quad , \quad \text{for } \omega < 100 \tag{6.62}$$

Therefore for frequency ω much lesser than 100 rad/s, the second term is approximately zero:

$$20\log_{10} \sqrt{\left(\frac{\omega}{100}\right)^2 + 1} \approx 20\log_{10} 1 = 0 \tag{6.63}$$

With the frequency dependent term of the magnitude approximately equal to zero for low frequencies much lesser than 100 rad/s, the magnitude is approximately equal to the first term:

$$20 \log_{10} |H(j\omega)| \approx 20 \log_{10} 10 = 20 \tag{6.64}$$

If we examine the magnitude plot in Fig. 6.2, it is quite easy to notice that for frequency less than around 20 rad/s, the magnitude is a constant value of 20 dB. This frequency of around 20 rad/s, correlates with our discussion above for the magnitude of the transfer function being approximately equal to the first term that is independent of frequency as the second term is negligible.

As the frequency increases past this approximate value of 20 rad/s, the magnitude is seen to decrease. Using a similar argument as done above, as frequency increases past 100 rad/s, the ratio $\frac{\omega}{100}$ is greater than 1 and $\left(\frac{\omega}{100}\right)^2 \gg 1$. This approximation results in the second term being simplified as follows:

$$20 \log_{10} \sqrt{\left(\frac{\omega}{100}\right)^2 + 1} = 20 \log_{10} \left(\frac{\omega}{100}\right) \tag{6.65}$$

With this, the second term has now been reduced to the logarithm of the ratio $\frac{\omega}{100}$. To understand the significance of this ratio, let us revisit the reason for choosing the x-axis of the Bode plot as the logarithm of the frequency. If the ratio:

$$\frac{\omega}{100} = 10^k \quad , \quad \text{for } k = 1, 2, 3, \ldots \tag{6.66}$$

The logarithm:

$$\log_{10} \left(\frac{\omega}{100}\right) = k \tag{6.67}$$

This might seem a little confusing at first. But (6.67) has converted the second term into a linear relationship with respect to the x-axis of the magnitude plot. The x-axis of the magnitude plot being $\log_{10} \omega$ appears in (6.67) in the form $\log_{10} \frac{\omega}{100}$. The ratio $\frac{\omega}{100}$ has a special interpretation. As this ratio changes from the power raised to one integer to the power raised to the next consecutive integer—from 1 to 2 to 3 etc.—the frequency is said to change by decades. A decade is when the frequency increases by a factor of 10. Therefore, with every decade increase in the frequency, the second term decreases by 20 dB. This can be formally expressed as

$$20 \log_{10} \left(\frac{\omega}{100}\right) = 20k \quad , \quad \text{for } k = \frac{\omega}{100} = 1, 2, 3, .. \tag{6.68}$$

In the above equation, two important points are to be noted. The expression is valid only for frequency $\omega \gg 100$ rad/s. And the linear relationship is valid as frequency increases as multiples of this frequency 100 rad/s.

With this discussion, if we examine the magnitude plot of Fig. 6.2, we find the magnitude decreasing linearly as the frequency is much greater than 100 rad/s. The reader is encouraged to verify by zooming in on the plots that the rate of change is indeed 20 dB per decade as the ratio $\frac{\omega}{100}$ increases. However, the magnitude plot is seen to deviate from the constant value of 20 dB after around 20 rad/s. This is because we have made an approximation for $\frac{\omega}{100}$ causing the second term to be negligible for $\omega \ll 100$ rad/s and for the second term to be linear for $\omega \gg 100$ rad/s. However, there will be a range of frequencies for which neither approximation is valid and this is where the magnitude is seen to have a non-linear relationship with the logarithm of the frequency.

The phase angle plot can also be interpreted in a similar manner. Before the widespread use of software for signal processing and control analysis, it was customary to draw magnitude plots and phase angle plots with pencil and paper. When doing so, the above approximations were very popular in being able to rapidly draw frequency response plots for transfer functions. However, now with readily available functions in almost every scientific computing software, it is no longer necessary to draw frequency response plots by hand. The phase angle plot can also be approximated to linear sections the way we did for the magnitude plot; however, this will not be necessary and will not be covered. The linear approximation of the magnitude plot on the other hand has a very important application in filter design as will be discussed in the coming sections and also will be used heavily in the next chapter.

6.9 Frequency Response of an LC Filter

In the previous section, we generated the Bode plots for the inductor and had interpreted the frequency response with respect to the transfer function. In this section, let us repeat the process for an LC filter by plotting the frequency response of the transfer function between the input voltage and the output voltage. As before, we will be using the bode function in the signal package available with Scipy.

In the previous chapter, we had derived the transfer function from the input voltage v_{in} to the output voltage v_o. The transfer function in the s domain is

$$\frac{V_o(s)}{V_{in}(s)} = H(s) = \frac{\frac{1}{LC}}{s^2 + \frac{R}{L}s + \frac{1}{LC}} \tag{6.69}$$

In the previous chapter, we had converted the transfer function from the continuous s domain to the digital z domain using Bilinear Transformation. In this section, we will not perform this conversion as in this section we will examine only the frequency response. In the next chapter, we will examine a simple function to convert a transfer function from the continuous s domain to the digital z domain. The frequency response on the LC filter will be significantly different from the inductor filter due to the second order term in the denominator.

The method of generating the Bode plots is exactly the same as the previous section. As we have already described the lti function and the bode function in the signal package in the previous section, we can just list the code for the LC filter:

```python
import numpy as np
import matplotlib.pyplot as plt
from scipy import signal
R = 0.1
C = 100.0e-6
L = 0.001
num_lc = [1/(L*C)]
den_lc = [1, R/L, 1/(L*C)]
lc_h = signal.lti(num_lc, den_lc)
w, mag, phase = signal.bode(lc_h, np.arange(100000))
plt.figure()
plt.title(''Magnitude'')
plt.xlabel(''Frequency'')
plt.ylabel(''dB'')
plt.grid()
plt.semilogx(w, mag)
plt.figure()
plt.title(''Phase angle'')
plt.xlabel(''Frequency'')
plt.ylabel(''Degrees'')
plt.grid()
plt.semilogx(w, phase)
plt.show()
```

The Bode plots generated by the above code are shown in Fig. 6.3

The frequency response plots are fairly different from the case of the simple inductor in the previous section. To interpret the frequency response especially for the magnitude plot, we can follow a very similar analysis as we did in the previous section. The magnitude plot has three distinct zones. The first zone is the low frequency zone where the magnitude appears to be constant. The second zone would be the high frequency zone where the magnitude decreases linearly with increasing frequency. These two zones are similar to the case of the inductor in the previous section. The third zone is the frequency range between the low frequency and the high frequency range, and in this zone the magnitude seems to have a spike. This spike in the magnitude is the resonant peak of the LC filter.

Let us rearrange the LC filter transfer function as follows:

$$H(s) = \frac{1}{LCs^2 + RCs + 1} \tag{6.70}$$

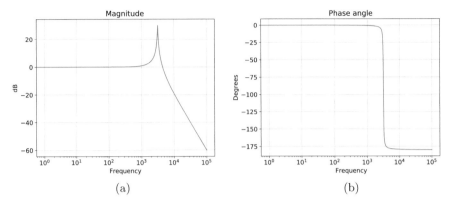

Fig. 6.3 Bode plots of the LC filter transfer function. (**a**) Magnitude plot (**b**) Phase angle plot

Expressing it with respect to frequency $s = j\omega$:

$$H(j\omega) = \frac{1}{-LC\omega^2 + j\omega RC + 1} \tag{6.71}$$

In (6.71), the term LC has a special significance with respect to the LC filter circuit. This term is related to the resonant frequency of the filter:

$$\omega_n = \frac{1}{\sqrt{LC}} \tag{6.72}$$

A circuit comprising of an inductor and a capacitor will experience resonance or oscillations. The frequency of these oscillations is inherent to the circuit and need not be externally supplied. For example, if a dc voltage were applied to an LC filter, there would oscillations in the output voltage before the output attained the steady state value equal to the input. The frequency of these oscillations is the resonant frequency of the LC filter.

To examine the first zone which is the low frequency zone, the frequency ω is lesser than the resonant frequency. Therefore:

$$\omega < \frac{1}{\sqrt{LC}} \tag{6.73}$$

which implies

$$\omega^2 << \frac{1}{LC} \tag{6.74}$$

This argument is similar to the previous section. In the low frequency zone, the LC filter transfer function is approximated to

$$H(j\omega) \approx \frac{1}{j\omega RC + 1} \tag{6.75}$$

For most practical LC filters, the product RC is usually small since the resistor R is merely the damping resistor and the filter capacitor C is usually much less than 1 for most practical applications. With this, the magnitude of the transfer function for the low frequency zone can be approximated to unity or 0 dB. A quick look at Fig. 6.3 shows that this is indeed the case.

The second zone is the high frequency zone where the frequency is greater than the resonant frequency. Using our usual square law:

$$\omega^2 >> \frac{1}{LC} \tag{6.76}$$

Using the above approximation in (6.71), the term $LC\omega^2$ is much greater than the other two terms RC and 1. The magnitude of the transfer function can be approximated to

$$|H(j\omega)| \approx \frac{1}{LC\omega^2} \tag{6.77}$$

Due to the presence of the second order term, the magnitude in decibels now has a different slope:

$$20 \log_{10} |H(j\omega)| \approx -20 \log_{10} LC - 40 \log_{10} \omega \tag{6.78}$$

The first term is a constant. However, the second term results in the magnitude changing by -40 decibels for every decade increase in the frequency ω. We have used the power rule of logarithms:

$$\log_{10} a^b = b \log_{10} a \tag{6.79}$$

From Fig. 6.3, the magnitude plot is seen to have a rate of change of -40 dB/decade in the high frequency zone. The reader in encouraged to zoom in on the high frequency range of the plot and verify this.

The third zone is the resonant peak. This zone occurs where the frequency is close to the resonant peak of the LC filter - $\omega \approx \frac{1}{\sqrt{LC}}$. Using this approximation in (6.71):

$$H(j\omega) \approx \frac{1}{j\omega RC} \tag{6.80}$$

As already stated before, the resistance of the filter is usually small as it is either the parasitic resistance of the inductor or is a damping resistor. In most cases, a filter capacitor will not have a value larger than a few milli Henry. Therefore, the product RC is usually quite small. This usually results in a resonant peak where the magnitude of the transfer function spikes around the resonant frequency. From Fig. 6.3, this spike is evident. The reader is encouraged to verify that the spike occurs at the resonant frequency of the LC filter.

With this we have examined the frequency response characteristics of the inductor and the LC filter. The first was a first order transfer function while second was a second order transfer function. In this section and the previous, we matched the behaviour of the magnitude plot with respect to the transfer function for different values of frequency. However, these frequency response characteristics are an incredibly powerful way to determine how a transfer function will behave as a filter. We will examine the physical significance of the Bode plots in the next section.

6.10   Physical Significance of these Frequency Responses

In the previous two sections, we examined the frequency response characteristics of the transfer functions of the inductor and of the LC filter. However, to be able to use frequency response characteristics as a tool to design filters, it is important to be able to interpret the frequency response characteristics. In this section, we will correlate the Bode plots of the LC filter with the physical behaviour of the filter.

In the previous chapter we have simulated the digital implementation of the LC filter transfer function and had examined the filtering action. We had seen how in some cases, the LC filter was able to remove the harmonic components present in the input and produce an output that consisted only of the fundamental. In some other cases, the LC filter was unable to remove the harmonics from the input as a result of which, the output was not a pure sinusoid. We can verify these results from frequency response characteristics.

The first question—what does it mean when a filter blocks a particular frequency component or removes it from the output? The main purpose of any filter is let something that you want to pass through and to stop everything else that you do not want. A simple example would be that of a coffee filter. The objective of a coffee filter is to collect the coffee grains as you pour the coffee through it while the brew which is what you want to drink is free of coffee grains. In the case of the LC filter, we are applying a voltage at the input. In the real world, a voltage available as a sensor measurement or any other output of a system will have noise and many other unwanted frequency components. Usually, the signal that we are interested in is low frequency near the grid frequency of 50 or 60 Hz. Noise and unwanted harmonics are signals of higher frequency that pollute the required signal. Some of these can be very high frequency components in the kHz or MHz range. Some on the other hand can be mere multiples of the grid frequency—150 Hz (3rd harmonic), 250 Hz (5th harmonic), etc. These are produced because of the specific operation of some devices – laptop/mobile chargers, LED lights, etc. These harmonic components can be integer multiples of the grid frequency or even in some cases non-integer multiples (inter-harmonics).

The LC filter will therefore accept as an input a voltage that could contain besides the fundamental grid frequency any of the abovementioned harmonics. The objective of the LC filter is to produce an output voltage that should retain some of the components in the input and remove the remaining components. Which are those some components? This depends on the application. In some applications, we would like to remove every component and leave only the fundamental. In some other applications, we might like to remove high frequency components in the kHz and MHz range and leave the lower frequency harmonics such as the 150, 250 Hz, etc.

When the filter is expected to remove a frequency component that is present in the input, how should the frequency response characteristic be like? If the magnitude plot of the Bode plots has a very low value for that frequency, that component will be almost eliminated from the output. For example, if there is a component of frequency ω_1 which is let us suppose for the sake of an example 5000 rad/s:

$$|Y(j\omega_1)| = |H(j\omega_1)||U(j\omega_1)| \qquad (6.81)$$

where Y is the output and U is the input. From Fig. 6.3, for frequencies greater than the resonant frequency, the magnitude of the LC filter transfer function decreases at the rate of 40 dB per decade. This implies that as the frequency increases, the magnitude of the transfer function will continue to decrease. Therefore, the filtering action of the LC filter will improve as the frequency increases.

The next question—when a filter is expected to let a component pass through, what should be the frequency response characteristic like? When a filter is expected to let a component pass through, the component should appear at the output with the minimal amount of change. Zero change is impossible—when applying any process to an input, the output will change even if that was not the intention. The objective is to minimize this change. There are now two parts of the Bode plot that matter— the magnitude and the phase angle plot. For the frequency of interest, the magnitude plot should have a value as close as possible to 0 dB. This is because 0 dB magnitude corresponds to an absolute gain of unity. Therefore, the magnitude of the component in the output will experience the minimal possible change with respect to the input. For the frequency of interest, the phase angle plot should have a value as close to 0 degrees as possible. As we already have seen before:

$$\arg(Y(j\omega_1)) = \arg(H(j\omega_1)) + \arg(U(j\omega_1)) \qquad (6.82)$$

If the phase angle deviates significantly from 0 degrees, the output will either lag behind the input or will lead the input. Both will be considered a distortion.

With the above two questions answered, let us sum up what we would expect from the LC filter. The design of the LC filter involves choosing a filter capacitor and the filter inductor. These two fix the resonant frequency of the LC filter. For frequencies lower than the resonant frequency, the magnitude of the Bode plot is close to 0 dB and therefore, these frequencies will be passed through with negligible amplification or attenuation. Moreover, the phase angle of the Bode plot is also close to 0 degrees, resulting in components significantly lower than the resonant frequency having negligible phase angle advance or delay. For frequencies much greater than the resonant frequency, the magnitude decreases linearly at the rate of 40 dB per decade. After a particular frequency, the magnitude will be so low as to render those frequency components to be negligible in the output. Therefore, for effective filter design, the choice of the resonant frequency is extremely important.

In this section, we have used frequency response characteristics to interpret how a filter would behave. We will revisit this concept in the next chapter, where we will further lay down design rules for designing a filter. In the next few sections, we will introduce a few more general concepts for digital filter design that we can use in the next chapter.

6.11 Generalized Poles and Zeros

In the past few sections, we examined the frequency response of the inductor transfer function and the LC filter transfer function. The main objective behind these sections was to add some analytical framework to the simulations presented in the previous chapter. However, a question that might linger for most readers is—why are we only talking about analog circuits? If we have to implement a filter digitally, do we always have to imagine the analog equivalent circuit, derive the transfer function, convert to the digital domain and then implement? This seems a roundabout method to implement filters and also restricts the capabilities of digital signal processing. In this section, we will talk about generalized transfer functions and how they can be used as building blocks for designing filters.

A generalized transfer function is a transfer function in the s domain that has a well-known structure due to which it has immediately inferred behaviour. There are many such generalized transfer functions—some of them simple and some of them very complex [4, 5]. It is important to remember that signal processing has been a topic that has been well researched for more than a century due to which significant contributions have been made by leading scientists all over the world. In this book, we will examine only some of the simple generalized transfer functions with the objective of using them in filter design in the next chapter.

6.11.1 The First Order Pole

The transfer function of a generalized first order pole is

$$H(s) = K \frac{1}{sT + 1} \tag{6.83}$$

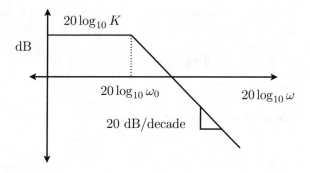

Fig. 6.4 Magnitude plot of a first order pole

Another very popular form to write this transfer function is

$$H(s) = K \frac{\omega_0}{s + \omega_0} \tag{6.84}$$

It is quite obvious that these two transfer functions are just different ways to express it and $\omega_0 = \frac{1}{T}$.

A quick description of the terms in the transfer function. ω_0 is the corner frequency or cut-off frequency, K is the gain and T is the time constant of the transfer function. At first sight, the transfer function of an inductor might seem something easier to relate to as being electrical engineers, we are at least comfortable with the presence of an inductor. But, we will soon see that the generalized first order pole has all the features of an inductor and actually provides more flexibility. Figure 6.4 shows a magnitude plot of a first order pole with respect to the terms in the transfer function. It is possible to generate an approximate phase angle plot as well; however, we will use the Python commands for that instead of using an analytical technique.

The gain K results in an offset to the gain of the entire transfer function. If $K > 1$, the entire magnitude plot shifts upwards and if $K < 1$, the plot shifts downwards. The magnitude of the transfer function remains constant at $20 \log_{10} K$ for frequencies less than the cut-off frequency ω_0. As frequency increases beyond ω_0, the magnitude of the transfer function decreases at the rate of 20 dB per decade. From Fig. 6.4 and from (6.84), we now have a very convenient and flexible way to design a basic filter using just two parameters—the cut-off frequency and the gain.

Let us demonstrate the use of this generalized first order pole transfer function with an example. Suppose, we want a basic filter that allows signals having a frequency less than 500 Hz to pass through to the output. On the other hand signals having a frequency greater than 500 Hz should experience some attenuation. At present, we will not set targets of attenuation for a particular frequency as this will be dealt with in detail in the next chapter on filter design. In this section, we are merely going to examine how to use the generalized transfer functions. We now have two specifications for our filter—the gain K should be unity and the cut-off frequency $\omega_0 = 2\pi \times 500 = 1000\pi$ rad/s. The cut-off frequency is quite simple to set once we know the frequency limit that needs to be passed through. The gain follows implicitly as when a signal needs to be passed through, the gain for that signal needs to be 1.

The code for the generalized first order pole is

```python
import numpy as np
import matplotlib.pyplot as plt
from scipy import signal
omega0 = 2*np.pi*500
K = 1
num_fop = [K*omega0]
den_fop = [1, omega0]
fop_tf = signal.lti(num_fop, den_fop)
w, mag, phase = signal.bode(fop_tf, np.arange(100000))
plt.figure()
plt.title(''Magnitude'')
plt.xlabel(''Frequency'')
plt.ylabel(''dB'')
plt.grid()
plt.semilogx(w, mag)
plt.figure()
plt.title(''Phase angle'')
plt.xlabel(''Frequency'')
plt.ylabel(''Degrees'')
plt.grid()
plt.semilogx(w, phase)
plt.show()
```

The Bode plots are shown in Fig. 6.5.

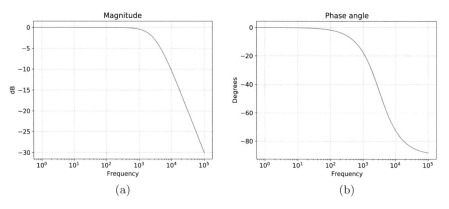

(a)   (b)

Fig. 6.5 Bode plots of the first order pole transfer function. (**a**) Magnitude plot (**b**) Phase angle plot

From the plots of Fig. 6.5, it is fairly obvious that they are very similar and equivalent to Fig. 6.2. The transfer function of a first order pole has therefore all the properties of a passive component such as an inductor and therefore has the same filtering properties as well. The advantage of using the first order filter instead of imagining an inductor is to not have to calculate values of inductance and the parasitic resistance in order to achieve a filter. For a filter, all we need to know is what needs to pass through and what needs to be blocked. Additionally, what is passed through can be amplified or attenuated and the first order pole has the gain factor that allows for that feature as well. Therefore, it is much easier to use the first order pole transfer function in filter design as will be done in the next chapter.

6.11.2   The Generalized Second Order Pole

In the same way as the generalized first order pole can be used to replace the transfer function of a passive component like an inductor, we can have a generalized second order pole that replaces the transfer function of an LC filter. The transfer function of a generalized transfer function would be

$$H(s) = K \frac{\omega_n^2}{s^2 + 2\zeta \omega_n s + \omega_n^2} \tag{6.85}$$

The above transfer function has three parameters or properties—gain K, resonant frequency ω_n and the damping ζ. Even though we are not talking about inductors and capacitors, a second order transfer function typically has a resonant frequency ω_n just like an inductor–capacitor circuit has. The damping ζ plays a similar role as compared to the resistor in the LC filter and affects the resonant peak. The gain K moves the entire magnitude plot upwards or downwards depending on its value being greater than or lesser than 1.

The code for the generalized first order pole is

```
import numpy as np
import matplotlib.pyplot as plt
from scipy import signal
omegan = 2*np.pi*500
K = 1
zeta = 0.1
num_sop = [K*omegan*omegan]
den_sop = [1, 2*zeta*omegan, omegan*omegan]
sop_h = signal.lti(num_sop, den_sop)
w, mag, phase = signal.bode(sop_h, np.arange(100000))
plt.figure()
plt.title(''Magnitude")
plt.xlabel(''Frequency")
```

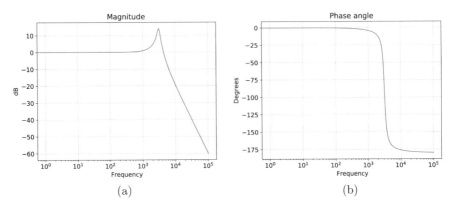

Fig. 6.6 Bode plots of the second order pole transfer function. (**a**) Magnitude plot (**b**) Phase angle plot

```
plt.ylabel(''dB")
plt.grid()
plt.semilogx(w, mag)
plt.figure()
plt.title(''Phase angle")
plt.xlabel(''Frequency")
plt.ylabel(''Degrees")
plt.grid()
plt.semilogx(w, phase)
plt.show()
```

The Bode plots are shown in Fig. 6.6.

The Bode plots of Fig. 6.6 are very similar to the Bode plots of Fig. 6.3. The generalized second order pole allows us to design a transfer function with the same properties as an LC filter but without designing equivalent circuit components. The resonant frequency ω_n determines the frequency beyond which the magnitude of the transfer function decreases at the rate to 40 dB per decade. The damping ζ determines the resonant peak—smaller values of ζ result in large resonant peaks while large values of ζ can completely remove the peak resulting in an over-damped second order transfer function.

6.11.3 Generalized First Order Zero

We could also synthesize transfer function with polynomials in the numerator. The transfer function of a first order zero is

$$H(s) = K \frac{s + \omega_0}{\omega_0} \tag{6.86}$$

It has the same two properties as a first order pole. Except that in terms of behaviour with respect to frequency, the first order zero behaves exactly the opposite of the first order pole.

The code for the first order zero is

```python
import numpy as np
import matplotlib.pyplot as plt
from scipy import signal
omega0 = 2*np.pi*500
K = 1
num_foz = [K, K*omega0]
den_foz = [omega0]
foz_tf = signal.lti(num_foz, den_foz)
w, mag, phase = signal.bode(foz_tf, np.arange(100000))
plt.figure()
plt.title(''Magnitude'')
plt.xlabel(''Frequency'')
plt.ylabel(''dB'')
plt.grid()
plt.semilogx(w, mag)
plt.figure()
plt.title(''Phase angle'')
plt.xlabel(''Frequency'')
plt.ylabel(''Degrees'')
plt.grid()
plt.semilogx(w, phase)
plt.show()
```

The Bode plots are shown in Fig. 6.7.

From a comparison of Figs. 6.7 and 6.5, it is quite obvious that they are exactly the opposite. The magnitude of the first order zero remains at the gain K until the cut-off frequency ω_0 and for frequencies greater than ω_0, the gain increases linearly at the rate of 20 dB per decade.

6.11.4 Generalized Second Order Zero

To conclude our array of transfer function building blocks, the next transfer function is the generalized second order zero:

$$H(s) = K \frac{s^2 + 2\zeta\omega_n s + \omega_n^2}{\omega_n^2} \tag{6.87}$$

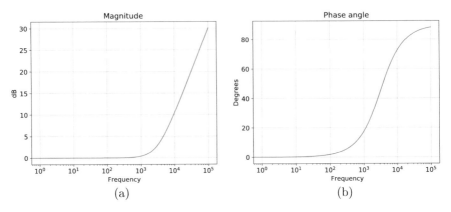

Fig. 6.7 Bode plots of the first order zero transfer function. (**a**) Magnitude plot (**b**) Phase angle plot

which is exactly the reciprocal of the generalized second order pole with the same parameters but with exactly the opposite behaviour with respect to frequency.

The code for the generalized second order pole is

```
import numpy as np
import matplotlib.pyplot as plt
from scipy import signal
omegan = 2*np.pi*500
K = 1
zeta = 0.1
num_soz = [K, K*2*zeta*omegan, K*omegan*omegan]
den_soz = [omegan*omegan]
soz_h = signal.lti(num_soz, den_soz)
w, mag, phase = signal.bode(soz_h, np.arange(100000))
plt.figure()
plt.title(''Magnitude'')
plt.xlabel(''Frequency'')
plt.ylabel(''dB'')
plt.grid()
plt.semilogx(w, mag)
plt.figure()
plt.title(''Phase angle'')
plt.xlabel(''Frequency'')
plt.ylabel(''Degrees'')
plt.grid()
plt.semilogx(w, phase)
plt.show()
```

The Bode plots are shown in Fig. 6.8.

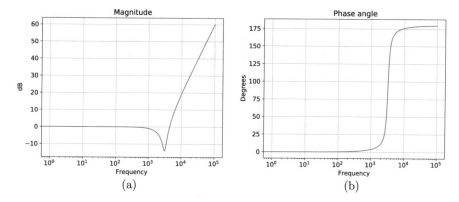

Fig. 6.8 Bode plots of the second order zero transfer function. (**a**) Magnitude plot (**b**) Phase angle plot

6.12   Conclusions

In this chapter, we have examined how we can study the behaviour of a transfer function with changing frequency. We had used Bode plots to study the frequency response characteristics of transfer functions. We had used the lti and the bode function in the signal package available with SciPy to define transfer functions and generate Bode plots using Python code. We had generated Bode plots for the inductor and the LC filter which we had simulated in the previous chapter. We had learned how to interpret the Bode plots to understand how a system behaves with changing frequency and we had interpreted the simulation results of the LC filter in the previous chapter.

We had started with generating Bode plots for the inductor and the LC filter so as to be able use our knowledge of analog circuits as filters. However, going further, we would approach digital filter design independent of analog circuits. For this purpose, we introduced the concept of generalized transfer functions which are transfer functions of a particular form having well-defined characteristics. We introduced the generalized first order pole, the second order pole, the first order zero and the second order zero. We had examined their frequency response characteristics and found that they are equivalent to analog circuits. In the next chapter, we will use these generalized transfer functions as building blocks in designing digital filters using Python code.

References

1. Brown, J. W., & Churchill, R. V. (1996). *Complex Variables and Applications* (6th ed.). McGraw-Hill.
2. Oppenheim, A. V., & Schafer, R. W. (2001). *Discrete-Time Signal Processing*. Pearson.

3. Van Valkenburg, M. (1984). In memoriam: Hendrik w. bode (1905–1982). *IEEE Transactions on Automatic Control, 29*(3), 193–194.
4. Conway, J. B. (1986). *Functions of One Complex Variable* (Vol. 1). Springer.
5. Henrici, P. (1974). *Applied and Computational Complex Analysis* (Vol. 1). Wiley.

Chapter 7
Filter Design

7.1 Introduction

In the previous chapter, we had described how frequency response characteristics can be examined using Bode plots for any given transfer function. We had also examined the physical significance of the Bode plots and how that relates to the way signals are transformed by a transfer function. We had introduced generalized transfer functions that can be synthesized without any reference to analog circuits but have similar properties to the transfer functions of analog circuits. We had stated that these generalized transfer functions are the building blocks for synthesizing higher order transfer functions.

In this chapter, we will design filters using the knowledge gained in the previous chapter. We will use frequency response characteristics to examine how transfer functions can behave as filters. We will consider two filter examples—the first being a low pass filter and the second being a band stop or notch filter. We will design our filters incrementally using the generalized transfer functions studied in the previous chapter. At every stage, Bode plots will be the tool used to determine the effectiveness of a particular transfer function as a filter.

This chapter will introduce new functions with SciPy and NumPy to help us design filters and implement them. We will examine how we can convert a transfer function from the continuous s domain into the digital z domain with the to_discrete method. This eliminates the need to manually convert a transfer function using the Bilinear Transformation. We will also examine how the result of the to_discrete method can be used to implement a filter in a simulation to verify its performance. We will also examine the polymul function available with NumPy to synthesize higher order transfer functions from the generalized transfer functions so as to avoid performing manual calculations.

This section will contain several code samples for quick reference along with the simulation results to prove concepts. Readers are encouraged to change code

S. V. Iyer, *Digital Filter Design using Python for Power Engineering Applications*,
https://doi.org/10.1007/978-3-030-61860-5_7

and experiment with values to examine the working of Python commands. This chapter will present effective tools for any engineer to design filters by combining basic signal processing and computational tools available with NumPy, SciPy and Python.

7.2   Designing Filters for Power Applications

Designing a filter usually follows similar steps no matter what the application is. However, designing filters for power applications is a bit simpler than other applications such as communications, computer vision or robotics. Applications such as voice or video communications need filters that are fairly advanced as the signals have frequencies over a wide range. To ensure noise free communication, specific signals need to be selected, while all others need to be rejected. For this reason, the filters used in communication tend to be fairly advanced. Though there is no rule prohibiting the application of these advanced filters in power engineering, it is like using a hammer to kill an ant. In this chapter, we will examine a simple method of synthesizing filters for power applications.

Even within the domain of power engineering, it is not possible to generalize the requirements of filters as the requirements from filters vary with the specific application. However, a few common cases are worth discussing in detail [1–3]. The most common filter is the low pass filter. As an example, when a voltage or a current is measured by a sensor, the signal available at the output will contain the actual voltage and will also contain noise. Noise can have a fairly wide range depending on the environment—can be in the kHz range or the MHz range. The simplest low pass filter that is usually implemented before feeding the signal to a controller is to remove the noise from the signal.

Besides removing the noise from a sensor output, a more advanced low pass filter involves removing harmonics from the signal. The fundamental frequency in power engineering is usually 50 or 60 Hz, while more specialized applications such as aerospace use 400 Hz. Besides the fundamental frequency, in most modern systems, there will be harmonics that are usually multiples of the fundamental frequency such as 150, 250 Hz, etc. These harmonics appear due to the operation of equipment that draw currents with harmonics even when the voltage supplied is a pure fundamental. As a specific example, if the sensor measures the voltage at some bus in the system, and this voltage contains harmonics besides the fundamental, we might need to extract the fundamental component. This would need a low pass filter that allows the fundamental to pass through but blocks all higher frequency components.

Another example could be thought of with such a voltage with harmonic components. In contrast to the low pass filter, we could imagine a case where we would like to extract the harmonics instead of the fundamental. Additionally, we might need to extract harmonics that have a frequency greater than a particular frequency. For example, we might identify frequencies above 1000 Hz as particularly problematic due to which we need to extract them and process them. In such a case, we need to implement a high pass filter, which will block all signals having a frequency less than 1000 Hz and pass through signals with a frequency greater than 1000 Hz. This application is quite common and is used for power quality assessment where the distribution company wishes to know the levels of harmonic pollution in their system.

Since we have talked about a low pass filter and a high pass filter, there is another filter that is between these two—the band stop filter or the notch filter. The notch filter gets its name from the shape of the magnitude plot in the frequency response characteristics. The notch filter will block a particular frequency range, thereby rejecting all signals in that frequency range. On the other hand, the notch filter will allow all other frequencies to pass through. An example of a notch filter would be when we identify a particular harmonic to be a problem creator. A very common culprit in power engineering is the triplen harmonics—150, 300, 450 Hz ($3f$). Sometimes, we would like to get rid of all of them or some of them before we process the signal. In such a case, we implement a notch filter with a frequency range around the frequency, which we wish to reject—for example, 150 Hz.

These are just a few examples, but in practice filtering requirements can be fairly complex with special applications needing filters that may not fall into a particular category. This brief description was only to provide some context to the two case studies that will be examined in detail in this chapter.

7.3 Getting Started with Low Pass Filter Design

Let us begin with our filter design examples with the case of a low pass filter. Let us define the requirements of the filter as follows. We will choose the fundamental grid frequency as 50 Hz. As already stated before, any voltage will have besides the fundamental, several harmonic frequency components. For the purpose of our filter design, let us define our objective as wishing to block all harmonic components equal to and greater than the 11th harmonic (550 Hz). This is just an example, but the process followed will help the reader design a low pass filter for any other purpose.

We will be using the approach of design, followed by verification, and will progress step by step until we arrive at a design that fulfils our requirements. We will use the frequency response characteristics in Bode plots as the design process and will use simulation for verification. We will introduce other commands available in Scipy's signal package and also packages with NumPy for both design and verification. We will use the generalized transfer functions described in the previous chapter as building blocks for our filters. We will introduce the concept of cascading transfer functions to fine-tune the filter design. In the previous chapter, we had briefly described how a filter should behave for frequency components that need to be passed through and for those other frequency components that need to be blocked. In this chapter, we will revisit this concept and start providing numerical goalposts to our filter requirements.

We could start our filter design with a rigorous mathematical approach. However, the advantage of having advanced scientific computing tools like Python and SciPy is the convenience of using coding to alter designs rapidly and repeatedly. Therefore, let us start our filter design with a hypothesis and gradually improve it. A safe start to any low pass filter design is with a generalized second order pole. The reason for this initial choice is that a second order pole has a fairly rapid decrease of 40 dB per decade in its magnitude for frequencies greater than the resonant frequency. Moreover, as will be shown soon, the second order pole is fairly easy to implement and is quite stable as opposed to filters with transfer function having higher orders.

As described in the previous chapter, the transfer function of the generalized second order pole is

$$H(s) = K \frac{\omega_n^2}{s^2 + 2\zeta\omega_n s + \omega_n^2} \tag{7.1}$$

To design a filter using a transfer function such as the one above, we need to choose values for the parameters which in this case are K, ζ and ω_n. Let us start with the gain K. Since we are designing a filter, we need to pass through components having a frequency less than the resonant frequency. Therefore, for these frequencies, the magnitude of the transfer function needs to be 1 or 0 dB. We can conveniently set the gain $K = 1$ as we do not need to amplify or attenuate the signals that need to be passed through.

Next parameter is ω_n. To choose ω_n, we need to use our knowledge of the magnitude of the transfer function decreasing at the rate of 40 dB per decade for frequencies higher than the resonant frequency. As the magnitude of the transfer function decreases, signals of increasing frequency are attenuated to a greater extent. To ensure that we are able to attenuate the 550 Hz component sufficiently, ω_n should be chosen sufficiently lower than 550 Hz. The next question is—is there a lower limit to how we can choose ω_n? We do not want the resonant peak that will appear at ω_n to be close to any frequency component that may appear and definitely greater than the fundamental 50 Hz as that is a component that we need to pass through. So

$$2\pi \times 50 \ll \omega_n \ll 2\pi \times 550 \tag{7.2}$$

The above inequality gives a quick glimpse into one of the challenges in filter design—having to compromise between signals you want to pass through and those you want to block. Soon, we will delve deeper into this.

The last parameter is ζ. As stated in the previous chapter, ζ is the damping applied to the resonance that occurs in a second order system. This is equivalent to the damping introduced by the resistor in the LC filter. Choosing ζ is a bit trickier. A system is said to be underdamped when ζ has a very small value due to which there is a large resonant peak. A large resonant peak is not a good sign, as this implies that the transfer function will have a large gain for any component with a frequency close to the resonant frequency. Besides the fundamental and the harmonics, we cannot completely rule out random frequencies that may appear in the system due to the operation of machines and equipment that may draw currents polluted with a lot of harmonics. Therefore, the accidental case of a resonant peak of a large magnitude for a frequency that may for some reason appear will result in this frequency component amplified in the output and disrupting further processing. However, one of the advantages of a sharp resonant peak is that the phase angle changes sharply at the resonant peak from $0°$ to $-180°$. Therefore, the fundamental frequency and other lower order harmonics do not suffer much phase angle delay.

On the other hand, an overdamped system is when the value of ζ is very large such that the resonant peak is completely suppressed. This may appear to be an attractive option at first, since it eliminates the possibility that any random component with a frequency close to the resonant frequency is unduly amplified. However, it brings in another problem. In an overdamped system, phase angle changes from $0°$ to $-180°$ very gradually over a wide frequency range. As a result, the fundamental frequency component or some other lower order harmonics might suffer a phase angle delay.

It is best to support the above discussion with design and simulation. Let us begin our design with $K = 1, \omega_n = 2\pi \times 275$ and $\zeta = 0.1$, which results in the following filter design code:

```
import numpy as np
import matplotlib.pyplot as plt
from scipy import signal
K = 1
omegan = 2*np.pi*275.0
zeta = 0.1
num_2ndpole = [K*omegan*omegan]
den_2ndpole = [1, 2*zeta*omegan, omegan*omegan]
secondpole_h = signal.lti(num_2ndpole, den_2ndpole)
w, mag, phase = signal.bode(secondpole_h,
np.arange(100000))
```

The latter part of the code for generating plots has been skipped to keep the code blocks brief. Figure 7.1 shows the Bode plots of this second order pole transfer function.

Let us examine the performance of this filter and decide where we could improve. We wish to block signals with frequencies greater than or equal to 550 Hz. I would recommend that the reader expands the code above with the plotting commands from the previous chapter and zooms in on the magnitude plot close to the frequency $2\pi \times 550 = 3455$ rad/s. We find that the magnitude of the transfer function is approximately -9.5 dB. This corresponds to an absolute magnitude of 0.33. The implication of this value is that any signal of frequency 550 Hz will appear in the output at 33% of its value at the input. Though we are achieving attenuation, the question is—is 0.33 good enough? Let us talk about this soon.

For the fundamental frequency of 50 Hz, the magnitude plot has a value of approximately 0.28 dB, which corresponds to an absolute magnitude of 1.03. This implies that any signal of frequency equal to 50 Hz appears at the output 3%

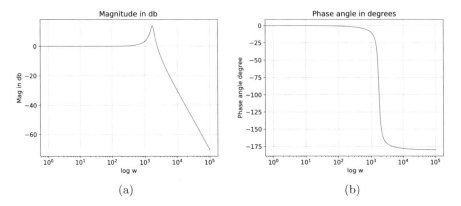

Fig. 7.1 Bode plots of the second order zero transfer function. (**a**) Magnitude plot. (**b**) Phase angle plot

greater than at the input. Since the fundamental frequency signal needs to be passed through by the filter, the phase angle of the transfer function at this frequency is also important. By zooming in on the phase angle plot close to the frequency $2\pi \times 50 = 314.1$ rad/s, the phase angle is found to be approximately $-2.1°$. This implies that any signal of frequency 50 Hz will have a phase delay of $-2.1°$ at the output with respect to the input.

The third feature of the transfer function is the resonant peak. The resonant peak has a peak magnitude of 14 dB, which corresponds to an absolute magnitude of 5. This implies that any signal of frequency close to the resonant frequency of 275 Hz will be amplified by a factor of 5 at the output with respect to the input. This method of assessing a filter's performance might seem heuristic; however, this method combines the convenience of using Python commands and graphical functionalities of software packages with insights that we have acquired from theory covered in the past chapter.

Let us now come around to assessing the performance of the filter. First question, is it acceptable that we attenuate a frequency that we wish to block to a value of 33% of its value at the input? Let us discuss what we expect from a filter when we wish to block a frequency. It is important to note that 100% removal of a frequency component is not physically possible. The requirement of removal is usually dictated by the application. In most engineering applications, if a signal of a particular frequency is attenuated to less than or equal to 5% of its value at the input, the signal is said to be effectively blocked or removed. Advanced applications such as aerospace or military applications can have more stringent requirements and in any case, while designing a filter, the requirements of the application are the final deciding factor. So, the attenuation of 33% that we have achieved with the second order pole would usually not be acceptable. Since we wish to remove the 550 Hz frequency, we do not need to think about the phase angle of the transfer function.

Next, let us discuss how effectively the filter passes through the fundamental frequency. With filter design, it is usually not possible to pass through a frequency without any modification. A filter will result in some non-zero attenuation or amplification and some non-zero phase angle delay or advance. This transfer function amplifies the fundamental 50 Hz frequency by 3%. The limits of amplification or attenuation for signals that need to be passed through by a filter are again decided by the application. However, a typical engineering application would expect the amplification or attenuation to be limited to a maximum of 5%. About the phase angle, the maximum phase angle delay or advance that can be tolerated for a 50 Hz sinusoid in most application is around 5–10°. With respect to these requirements, the second order pole does pass the acceptability criteria.

Now, the resonant peak results in a gain of 5 for frequencies close to 275 Hz. In most power engineering applications, it is quite common for a 5th or 250 Hz frequency component being present in a signal. A large resonant peak of 5 is a cause of concern as the 250 Hz component in the signal could be significantly amplified. Therefore, the second order pole would not be acceptable with such a large resonant peak. In conclusion, this first design of a low pass filter falls short on two criteria— removing high frequency components and a large resonant peak. Let us look at the other extreme of an overdamped transfer function to examine how the performance compares with the underdamped transfer function that has been examined.

Figure 7.2 shows the Bode plots for the second order pole with parameters $K = 1$, $\omega_n = 2\pi \times 275$ and $\zeta = 1.5$. Due to the value of the damping factor $\zeta = 1.5$, this transfer function is now overdamped. In the magnitude plot of Fig. 7.2, we see that the resonant peak has been completed suppressed. This is the equivalent of choosing such a large resistor in an LC filter so as to present a large resistance at the resonant frequency. From the discussion above of the underdamped second order pole, the absence of a resonant peak might seem like a solution as now we do not have to worry about a stray component with frequency close to the resonant

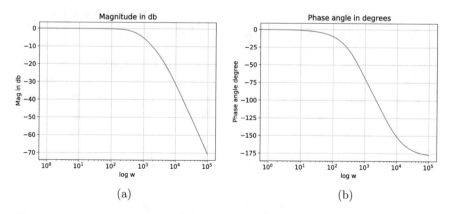

Fig. 7.2 Bode plots of the second order zero transfer function. (**a**) Magnitude plot. (**b**) Phase angle plot

frequency unduly amplified by the filter. However, a closer examination of the phase angle plot of Fig. 7.2 will reveal the problem. The phase angle gradually changes from 0 to $-180°$ over several decades. Most worryingly, at the frequency of $2\pi \times$ 50 rad/s corresponding to the fundamental grid frequency, the value of the phase angle is approximately $-29°$. This implies that the fundamental component will experience a phase delay of $-29°$ at the output with respect to the input. Such a large phase delay will almost always be unacceptable.

Comparing the underdamped and overdamped cases, it is evident that neither of them will be a suitable solution. We need to trade-off between the two frequency response plots. We would like the sharp cliff-type frequency response plots of the underdamped case but without the large resonant peak. The reader is encouraged to try out the second order filter design code with different values of damping factor ζ. The question that the filter designer needs to answer to achieve a reasonable trade-off is—how much of phase delay at fundamental frequency is acceptable and how much magnitude at the resonant frequency can the transfer function have? To answer this question, the filter designer might need to consult the application engineers who will design the system that follows the low pass filter.

Let us redesign the second order pole by changing the damping factor by making $\zeta = 0.3$. Figure 7.3 shows the Bode plots of the transfer function with this value of damping factor. It is important to note from the magnitude plot that the resonant peak has not been completely suppressed. However, the larger damping factor of 0.3 has decreased its peak significantly. The resonant peak has a maximum gain of approximately 5 dB, which corresponds to an absolute magnitude of 1.77. This implies that a signal of frequency 275 Hz will be magnified by a factor of 1.77. This is significantly lower than the underdamped case where the absolute magnitude was 5. There may, however, be applications where a magnitude of 1.77 is also unacceptable in which case, the only option would be to use an overdamped second order pole. Additionally, the value of the phase angle plot for the fundamental

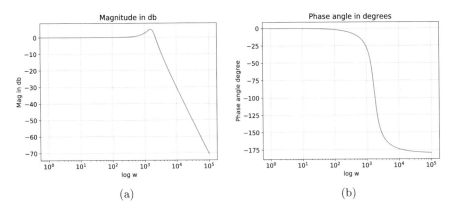

Fig. 7.3 Bode plots of the second order zero transfer function. (**a**) Magnitude plot. (**b**) Phase angle plot

frequency of 50 Hz is $-6°$, which is also much lesser than the overdamped case. It is quite obvious that this damping factor of 0.3 is indeed the middle ground between the underdamped transfer function and the overdamped transfer function. The reader is encouraged to zoom in on the magnitude plot for the frequency 550 Hz and verify that the magnitude of the transfer function is approximately -9 dB. This implies that even though we may decrease the resonant peak and decrease the phase angle delay at the fundamental frequency, the transfer function would not be able to sufficiently attenuate frequencies higher than 550 Hz.

As already stated before, this method of designing filters might seem heuristic. However, it is extremely convenient, given the vast array of scientific computing tools that are available to us. We have used frequency response characteristics to arrive at our first design of a low pass filter. The design is by no means complete as we still have not successfully removed signal components with frequencies greater than 550 Hz. Before improving on this design, let us simulate this transfer function and map the simulation results with the analytical results derived from the frequency response characteristics.

7.4 Simulating a Filter Transfer Function

In the chapter on analog filters, we simulated the digital implementation of the inductor, capacitor and the LC filter. Through the simulations, we had shown how we could begin with a circuit, derive a transfer function in the continuous frequency s, convert the transfer function into the digital form in the digital frequency z and finally arrive at a difference equation. This difference equation can be solved iteratively in a simulation or can be implemented in hardware using a microcontroller. We can repeat this process for the second order pole that we have synthesized in the previous section. There will be, however, a difference in our approach. In the analog filters chapter, we had used the Bilinear Transformation to derive the transfer functions in the digital frequency z. In this section, we will examine a pure programmatic approach to generating the transfer function in the digital z domain, which will make the process of filter design much simpler.

In the previous section, we had used the lti function with the signal package to define the transfer function. In the previous chapter, we had described how the lti function creates an object that contains data and several methods and attributes. The reader is encouraged to quickly review the concepts of Object Oriented Programming (OOP) described in the previous chapter as we will be using it in this section. Let us re-examine the statement that defines the transfer function:

```
secondpole_h = signal.lti(num_2ndpole, den_2ndpole)
```

The lti function produces an object that is assigned to secondpole_h. In the previous section, we had passed this object to the bode function to generate the data for the Bode plot. The object secondpole_h contains several methods and attributes that allow us to perform commonly needed computations and transformations. One of these methods is the to_discrete method.

The to_discrete method is available with any object produced by the lti function and converts the transfer function represented by the object into the discrete or digital form. With this method, we do not need to use the Bilinear Transformation and manually rearrange the resultant transfer function. The to_discrete method does all that for us thereby making the implementation of digital filters much simpler and less error prone. The usage of the to_discrete method is as follows:

```
secondpole_h.to_discrete(dt, method, alpha)
```

It is important to note that the to_discrete method needs to be with the object secondpole_h as it is a method of the object. This is in contrast to the bode function, which is a function within the signal package and is independent of the lti object.

The to_discrete method takes three arguments though only the first argument dt is a required argument, while the second and third arguments have default values. While converting a transfer function into the discrete or digital form, the most important factor is the sampling time interval in the final digital form. The argument dt allows us to specify this sampling time interval. The next most important aspect in the conversion process is the method used. In Chap. 3, we had covered the different methods that can be used to convert a transfer function from continuous into digital form. We had covered the zero order hold, the first order hold and the Bilinear Transformation. These can be specified in the to_discrete method with the method argument. The possible values for the argument can be zoh (zero order hold), foh (first order hold), euler (Euler's method) or tustin (Bilinear Transformation). If this argument is skipped, the default value is method=zoh. Since we would like to use the Bilinear Transformation due to its accuracy and stability, we will specify method='tustin'. The last argument alpha is a weighting parameter that is used with another method that is available called the Generalized Bilinear Transformation. However, the argument alpha can be safely skipped in which case its default value will be None.

The code written in the previous section can be extended as follows:

```
import numpy as np
import matplotlib.pyplot as plt
from scipy import signal
K = 1
omegan = 2*np.pi*275.0
zeta = 0.3
num_2ndpole = [K*omegan*omegan]
den_2ndpole = [1, 2*zeta*omegan, omegan*omegan]
secondpole_h = signal.lti(num_2ndpole, den_2ndpole)
w, mag, phase = signal.bode(
                        secondpole_h,
                        np.arange(100000)
                        )
secondpole_h_z = secondpole_h.to_discrete(
                        dt=200.0e−6,
```

$$method = 'tustin'$$
$$)$$

We have assigned the result of the to_discrete method to another variable to avoid overwriting our original continuous transfer function. We are using a 200 μs sampling time similar to what we used in the chapter on analog filters. The next question is—what will be second_pole_h_z?

The value of second_pole_h_z can be printed to the console as

```
TransferFunctionDiscrete(
array([0.02633861, 0.05267722, 0.02633861]),
array([ 1.        , -1.71172543,  0.81707986]),
dt: 0.0002
)
```

A comparison with the value of second_pole_h in the previous chapter makes it evident that the object is fairly similar. However, second_pole_h_z is of type TransferFunctionDiscrete as opposed to TransferFunctionContinuous. Transfer-FunctionDiscrete is another class constructor for transfer functions but in the digital or discrete domain and produces a transfer function object. The first argument is an array that represents the numerator polynomial, while the second argument is the array that represents the denominator polynomial. The third argument is the sampling time of 200 μs that was passed to the to_discrete method.

The most important feature of the to_discrete method is that it produces the final result of the transfer function in the digital domain. The above object is equivalent to the transfer function

$$H(z) = \frac{0.02633861z^2 + 0.05267722z + 0.02633861}{z^2 - 1.71172543z + 0.81707986} \qquad (7.3)$$

The third argument dt is merely an indication that the sampling time of 200 μs has been applied in the conversion. However, the TransferFunctionDiscrete produces a transfer function in the digital domain, and this can be used directly in our implementation.

Since the TransferFunctionDiscrete is a class, the object created from the class has methods and attributes. Two very convenient attributes are num and den, which provide the numerator and denominator polynomials as arrays. For example,

secondpole_h_z.num

produces

```
[0.02633861, 0.05267722, 0.02633861]
```

while

secondpole_h_z.den

produces

```
[ 1.        , -1.71172543,  0.81707986]
```

Therefore, we could express the difference equation with respect to the numerator
and the denominator of the resultant transfer function in the digital domain.

Let us illustrate how we can simulate a digital implementation of a transfer
function using the following code:

```python
import numpy as np
import matplotlib.pyplot as plt
from scipy import signal
K = 1
omegan = 2*np.pi*275.0
zeta = 0.3
num_2ndpole = [K*omegan*omegan]
den_2ndpole = [1, 2*zeta*omegan, omegan*omegan]
secondpole_h = signal.lti(num_2ndpole, den_2ndpole)
w, mag, phase = signal.bode(
                          secondpole_h,
                          np.arange(100000)
                          )
secondpole_h_z = secondpole_h.to_discrete(
                          dt=200.0e-6,
                          method='tustin'
                          )

t_duration = 1.0
t_step = 1.0e-6
no_of_data = int(t_duration/t_step)
time_array = np.arange(no_of_data)*t_step

frequency = 50.0
omega = 2*np.pi*frequency
omega_noise = 2*np.pi*550.0
inp_mag = np.sqrt(2)*240.0
ip_voltage_signal = (np.sin(time_array*omega) \
        + 0.3*np.sin(time_array*omega_noise))

t_sample = 200.0e-6
no_of_skip = int(t_sample/t_step)
tsample_array = time_array[::no_of_skip]
ip_voltage_samples = ip_voltage_signal[::no_of_skip]
op_voltage_samples = np.zeros(ip_voltage_samples.size)
u = np.zeros(3)
y = np.zeros(3)

for volt_index, volt_value in np.ndenumerate
(ip_voltage_samples):
```

```
u[0] = volt_value
y[0] = ( secondpole_h_z.num[0] * u[0] + \
secondpole_h_z.num[1] * u[1] + \
secondpole_h_z.num[2] * u[2] - \
secondpole_h_z.den[1] * y[1] - \
secondpole_h_z.den[2] * y[2] ) / 
secondpole_h_z.den[0]
u[2] = u[1]
y[2] = y[1]
u[1] = u[0]
y[1] = y[0]
op_voltage_samples[volt_index] = y[0]

plt.figure()
plt.plot(tsample_array, ip_voltage_samples, \
            label='input', ds='steps')
plt.plot(tsample_array, op_voltage_samples, \
            label='output', ds='steps')
plt.title('Input versus output')
plt.xlabel('Time')
plt.ylabel('Volts')
plt.legend()
plt.show()
```

The code block above is largely identical to the code presented in the Chap. 5, where we had simulated the digital implementation of an LC filter. The only difference lies in the fact that we have not used any formulas or expression but have merely used Python commands. The implementation of the digital filter as a difference equation uses the numerator and denominator arrays. A simple comparison between the implementation above and that in Chap. 5 will show that the general transfer function in the digital domain

$$H(z) = \frac{Y(z)}{U(z)} = \frac{n_2 z^2 + n_1 z + n_0}{d_2 z^2 + d_1 z + d_0} \qquad (7.4)$$

will be implemented as

$$y[n] = \frac{n_2 u[n] + n_1 u[n-1] + n_0 u[n-2] - d_1 y[n-1] - d_0 y[n-2]}{d_2} \qquad (7.5)$$

The result of the simulation is shown in Fig. 7.4. The plot has the input and output waveforms, and it can be seen that although the harmonic content in the output is much lesser than in the input, the output waveform is still quite distorted. Our desired result is for the output waveform to be as close as possible to a sinusoid of 50 Hz frequency with the minimal amount of higher order harmonics. This is due

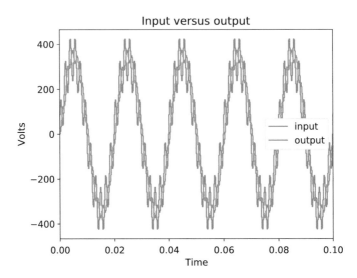

Fig. 7.4 Performance of the second order pole as filter

to the fact that in our code block above, we have considered the input to be a sum of a fundamental 50 Hz sinusoid with a 30% 550 Hz harmonic content. The presence of the 550 Hz component in the output can be attributed to the fact that the second order pole chosen as a filter achieves a 0.33 attenuation to the 550 Hz component as seen from the frequency response characteristics of Fig. 7.3.

In this section, we started off on our filter design journey with a second order pole. We had used Bode plots to examine the frequency response of the filter to determine if the performance of the filter would be adequate. We had also used Bode plots to fine-tune parameters of the filter to avoid potential amplification of signals due to the resonant nature of the second order pole. We had determined that the filter would probably not be effective in adequately attenuating the 550 Hz harmonic component. We had simulated the designed filter to confirm that the performance of the filter was indeed inadequate. In the next section, we will examine how the performance of the filter can be improved.

7.5   Cascading Transfer Functions

In the previous section, we started the filter design process with a second order pole and had found the performance of this filter to be inadequate. This would imply we need to choose another filter. Instead of completely rejecting this first design and opting for another advanced higher order filter, in this section, we will instead choose the approach of synthesizing a filter using transfer functions as building blocks. In the previous chapter, we had presented several generalized transfer functions of which we had used the second order pole. We will cascade transfer functions by using generalized transfer functions as building blocks to create higher order transfer functions.

Cascading transfer functions implies connecting the output of the first to the input of the second. The input signal is fed to the input of the first transfer function, and the output signal is available at the output of the second transfer function. Mathematically, in the frequency s domain, the combined transfer function is the product of the two transfer functions

$$H(s) = H_1(s)H_2(s) \qquad (7.6)$$

It is quite obvious that cascading transfer functions will result in a higher order transfer function as we are multiplying polynomials. However, as will be shown later, the implementation remains far simpler as each transfer function is implemented separately as they are connected together.

The only shortcoming of the second order pole as a filter was the insufficient attenuation at the minimal harmonic component 550 Hz. Therefore, let us choose the two transfer functions to be cascaded to be identical to the second order pole above

$$H_1(s) = H_2(s) = \frac{\omega_0^2}{s^2 + 2\zeta\omega_0 + \omega_0^2} \qquad (7.7)$$

with $\omega_0 = 2\pi \times 275$ rad/s and $\zeta = 0.3$. We will repeat the process followed in the previous section—frequency response analysis followed by simulation to determine the effectiveness of this filter. To begin with, we need to express the product of the two transfer functions $H_1(s)$ and $H_2(s)$. Though, in this particular case, we could just write the product by mental calculation, we would like to use a computational method that can be applied to more complex cases where paper–pencil calculations would be error prone.

To compute the product of the transfer functions, we need to introduce another function. But before that, let us examine the need for this function. We could define the two transfer functions using the lti function in the signal package as before

```
num1_2ndpole = [omega_0*omega_0]
den1_2ndpole = [1, 2*zeta*omega_0, omega_0*omega_0]
tf1_h = signal.lti(num1_2ndpole, den1_2ndpole)
num2_2ndpole = [omega_0*omega_0]
den2_2ndpole = [1, 2*zeta*omega_0, omega_0*omega_0]
tf2_h = signal.lti(num2_2ndpole, den2_2ndpole)
```

We have created two transfer function objects tf1_h and tf2_h. However, it is to be noted that these objects are Python objects. At the time of writing this book, we cannot multiply these objects together,

```
tf1_h * tf2_h
```

will produce this error:

```
TypeError: unsupported operand type(s) for *:
'TransferFunctionContinuous' and
'TransferFunctionContinuous'
```

The error implies that we cannot multiply two transfer functions together as the multiply (*) operator has not been defined for these objects. I would still encourage the reader to try out this operation as with time, someone might have added this feature.

It would have been very convenient if the multiply operator could have been used as we attempted to above. However, there is another very convenient manner to synthesize higher order transfer functions as products of lower order transfer functions. We use the polymul function available with NumPy to multiply the polynomials that are represented by arrays.

```
num_product = np.polymul(num1_2ndpole, num2_2ndpole)
den_product = np.polymul(den1_2ndpole, den2_2ndpole)
```

num1_2ndpole and num2_2ndpole are array representations of polynomials. We have fed these to the lti function to create transfer function objects. However, the polymul function with NumPy can multiply two such polynomials represented by arrays and produce an array representation of the polynomial product. The result

```
[8.3521e+12]
[1.0000e+00, 2.0400e+03, 6.8204e+06, 5.8956e+09,
 8.3521e+12]
```

The polymul can therefore be used to multiply polynomials so as to separately compute the numerator and denominator of the resultant transfer function. In the later section, when we synthesize a notch filter, we will find that this polymul function simplifies filter design significantly. Now that we have computed the numerator and denominator of the resultant transfer function product, we can use the lti function:

```
tf_product_h = signal.lti(num_product, den_product)
```

With the polymul function, we have avoided any manual computations and can write code that needs only the parameters of the generalized transfer function as inputs.

The entire code is

```
import numpy as np
import matplotlib.pyplot as plt
from scipy import signal

omega_0 = 2*np.pi*275.0
zeta = 0.3

num1_2ndpole = [omega_0*omega_0]
```

```
den1_2ndpole  =  [1,  2*zeta*omega_0,  omega_0*omega_0]
num2_2ndpole  =  [omega_0*omega_0]
den2_2ndpole  =  [1,  2*zeta*omega_0,  omega_0*omega_0]
num_product  =  np.polymul(num1_2ndpole,  num2_2ndpole)
den_product  =  np.polymul(den1_2ndpole,  den2_2ndpole)

tf_product_h  =  signal.lti(num_product,  den_product)
w,  mag,  phase  =  signal.bode(tf_product_h,
 np.arange(100000))

plt.figure()
plt.title('Magnitude in db')
plt.xlabel('log w')
plt.ylabel('Mag in db')
plt.grid()
plt.semilogx(w,  mag)

plt.figure()
plt.title('Phase angle in degrees')
plt.xlabel('log w')
plt.ylabel('Phase angle degree')
plt.grid()
plt.semilogx(w,  phase)

plt.show()
```

The frequency response characteristics are shown in Fig. 7.5.

Let us compare the frequency response of Fig. 7.5 with that of Fig. 7.3. The gain of the cascaded transfer function at 550 Hz is −21 dB, which corresponds to an absolute magnitude of 0.089. This is significantly lower than the magnitude of

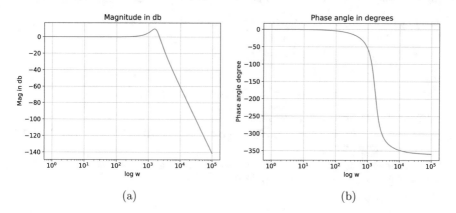

(a)   (b)

Fig. 7.5 Bode plots of the cascaded transfer function. (**a**) Magnitude plot. (**b**) Phase angle plot

0.33 achieved by a single second order pole. The phase angle delay of the cascaded transfer function at the fundamental 50 Hz frequency is $-13.1°$. This is greater than the phase angle delay of $-6°$ with a single second order pole. However, a phase angle delay of $-13.1°$ would be acceptable under most applications and would not result in a significant distortion. The magnitude at the resonant peak for the cascaded transfer function is found to be 10 dB, which corresponds to an absolute value of 3.16. Again, this magnitude is greater than the value of 1.77 realized with a simple second order pole. However, 3.16 might also be seen as an acceptable resonant peak.

From our analysis of the frequency response characteristics of the cascaded transfer function comprised of two second order poles, we find that we have achieved a trade-off. We have significantly improved the attenuation of higher order harmonics. However, we have increased the phase angle delay at the fundamental frequency and the magnitude of the resonant peak. At this point of time, it is advisable for the filter designer to consult with the application engineer whether these results are acceptable. In all filters, the pass/fail judgment lies with the application engineer. In the case of a fail, the filter designer needs to continue with the filter design process by either increasing the order of the filter or changing parameters. At this moment, we will stop the filter design and proceed to implementing this cascaded transfer function.

7.6   Simulating the Cascaded Low Pass Filter

In the previous section, we had demonstrated the use of the polymul function with NumPy before calling the lti function to synthesize a higher order transfer function using second order poles. It might seem an obvious step to convert the final transfer function into the digital domain using the to_discrete form as done before. However, there is a potential complication which we will discuss along with the solution.

To begin with, let us use the to_discrete method to convert the higher order transfer function from the continuous s domain into the digital z domain

```
tf_product_h_z = tf_product_h.to_discrete(
                 dt=200.0e-6,
                 method='tustin'
                 )
```

The resultant transfer function in the digital or discrete domain is

```
TransferFunctionDiscrete(
array([0.00065305, 0.00261221, 0.00391831, 0.00261221,
 0.00065305]),
array([ 1.        ,  -3.43478645,   4.58866489,
 -2.81519459,   0.67176497]),
dt: 0.0002
)
```

which would translate to the following transfer function:

$$H(z) = \frac{0.00065305z^4 + 0.00261221z^3 + 0.00391831z^2 + 0.00261221z + 0.00065305}{z^4 - 3.43478645z^3 + 4.58866489z^2 - 2.81519459z + 0.67176497}$$
(7.8)

At first glance, there does not seem to anything "wrong" with the above digital domain transfer function. The problem, however, is a little more subtle and can appear in hardware implementation rather than mere simulation. Upon closer look, the coefficients of the numerator polynomial are several degrees smaller than the coefficients of the denominator polynomial. In absolute magnitude, the largest and smallest coefficients are 4.58866489 and 0.00065305, respectively. When the coefficients of the polynomials vary as widely as they do above, when implementing the filter using a microcontroller, it is possible that the filter might be unstable due to resolution errors. This is due to the fact that the microcontroller might not be able to resolve the coefficients with reasonable accuracy. The resulting instability is quite often perplexing and difficult to debug. However, the greater the power of the continuous domain transfer function and the lower the sampling time interval, the greater are the chances of instability. This is due to the Bilinear Transformation used in the conversion process:

$$s = \frac{2}{T}\frac{z-1}{z+1}$$
(7.9)

There are a few methods to avoid implementation instability. One of the common reasons for instability is choosing a sampling time interval that is much lower than what might be needed. When choosing the sampling time interval, use the practical Nyqvist's criterion—the sampling frequency must be at least 10 times the frequency of the signal of interest. Choosing a sampling frequency that is only slightly higher than the system frequency will result in a distorted signal as the dynamics of the signal have not been fully captured. Choosing a sampling frequency that is arbitrarily larger than the system frequency (for example, 100,000 times) may seem like a solution. But, a very high sampling frequency (which implies a very low sampling time interval) will result in conversion problems for transfer functions designed in the continuous time domain. Therefore, avoid choosing a too large sampling frequency with respect to the system.

The other solution is to avoid implementing higher order transfer functions and instead implementing the lower order building blocks. We had implemented the second order pole in the past section. Instead of implementing the higher order transfer function, we could implement two second order poles. Since these second order poles are cascaded, the implementations can be cascaded as well—the output of the first can be fed to the input to the other. From an implementation standpoint, it may use more memory due to the need to store more coefficients and samples. However, with modern microcontrollers, memory is usually not a constraint as most modern DSPs have several megabytes of on-chip memory and it is only a few more coefficients and samples that we are adding.

The code for the implementation of the higher order filter as cascaded lower order filters is as follows:

```python
import numpy as np
import matplotlib.pyplot as plt
from scipy import signal

omega_1 = 2*np.pi*275.0
zeta1 = 0.3
omega_2 = 2*np.pi*275.0
zeta2 = 0.3
num1_2ndpole = [omega_1*omega_1]
den1_2ndpole = [1, 2*zeta1*omega_1, omega_1*omega_1]
secondpole_h1 = signal.lti(num1_2ndpole, den1_2ndpole)
num2_2ndpole = [omega_2*omega_2]
den2_2ndpole = [1, 2*zeta2*omega_2, omega_2*omega_2]
secondpole_h2 = signal.lti(num2_2ndpole, den2_2ndpole)

secondpole_h1_z = secondpole_h1.to_discrete(
        dt=200.0e-6,
        method='tustin'
        )
secondpole_h2_z = secondpole_h2.to_discrete(
        dt=200.0e-6,
        method='tustin'
        )

t_duration = 1.0
t_step = 1.0e-6
no_of_data = int(t_duration/t_step)
time_array = np.arange(no_of_data)*t_step

frequency = 50.0
omega = 2*np.pi*frequency
omega_noise = 2*np.pi*550.0
inp_mag = np.sqrt(2)*240.0
ip_voltage_signal = inp_mag*(
                np.sin(time_array*omega) + \
                0.3*np.sin(time_array*omega_noise)
                )
t_sample = 200.0e-6
no_of_skip = int(t_sample/t_step)
tsample_array = time_array[::no_of_skip]
ip_voltage_samples = ip_voltage_signal[::no_of_skip]
op_voltage_samples = np.zeros(ip_voltage_samples.size)
```

```python
u1 = np.zeros(3)
y1 = np.zeros(3)
u2 = np.zeros(3)
y2 = np.zeros(3)

for volt_index, volt_value in np.ndenumerate
(ip_voltage_samples):
    u1[0] = volt_value
    y1[0] = ( secondpole_h1_z.num[0] * u1[0] +\
        secondpole_h1_z.num[1] * u1[1] + \
        secondpole_h1_z.num[2] * u1[2] - \
        secondpole_h1_z.den[1] * y1[1] - \
        secondpole_h1_z.den[2] * y1[2] ) / 
        secondpole_h1_z.den[0]
    u1[2] = u1[1]
    y1[2] = y1[1]
    u1[1] = u1[0]
    y1[1] = y1[0]
    u2[0] = y1[0]
    y2[0] = ( secondpole_h2_z.num[0] * u2[0] + \
        secondpole_h2_z.num[1] * u2[1] + \
        secondpole_h2_z.num[2] * u2[2] - \
        secondpole_h2_z.den[1] * y2[1] - \
        secondpole_h2_z.den[2] * y2[2] ) / 
        secondpole_h2_z.den[0]
    u2[2] = u2[1]
    y2[2] = y2[1]
    u2[1] = u2[0]
    y2[1] = y2[0]
    op_voltage_samples[volt_index] = y2[0]

plt.figure()
plt.plot(tsample_array, ip_voltage_samples, \
        label='input', ds='steps')
plt.plot(tsample_array, op_voltage_samples, \
        label='output', ds='steps')
plt.title('Input versus output')
plt.xlabel('Time')
plt.ylabel('Volts')
plt.grid()
plt.legend()
plt.show()
```

The simulation results are shown in Fig. 7.6.

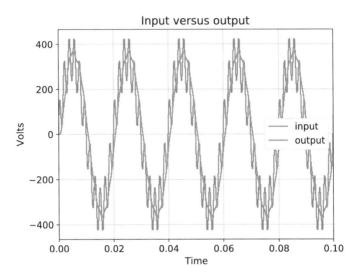

Fig. 7.6 Performance of the higher order transfer function

A comparison of the simulation results of Fig. 7.6 with those of Fig. 7.4 makes it evident that the filtering action of the higher order filter is much more effective. The harmonic content in the output has significantly decreased, and the output waveform is much closer to a sinusoid of fundamental 50 Hz frequency. Though the output waveform has lower harmonic content, it is important as before to check with the application engineer if the filtering action is acceptable. Advanced applications such as aerospace and military might have very stringent requirements that may need very low harmonic content in the output. In such cases, the filter designer will need to follow the same steps as in the past two sections in improving the filter design. However, with this design, we will stop with the low pass filter and examine the next case of a notch filter, which will be another interesting design example on cascading lower order transfer functions.

7.7   Getting Started with a Notch Filter

In the previous few sections, we had designed a low pass filter and verified its performance through simulating the digital implementation of the filter. We had used frequency response characteristics using Bode plots as a design tool. In this section, we will use the same technique of frequency response based design followed by simulations to design a notch filter. There are many different techniques to design filters for a number of different applications. However, the frequency response based design is a very simple and effective method of designing basic filters that can be translated fairly easily to a digital hardware implementation.

Fig. 7.7 Frequency response
of a notch filter

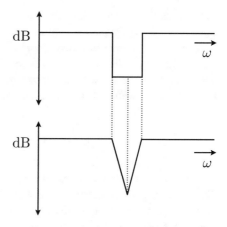

A notch filter gets its name from the shape of the magnitude of the frequency response characteristic. Another name for the notch filter is a band stop filter as this filter blocks a band of frequencies, while passing frequencies lower than the band as well as frequencies greater than the band. Figure 7.7 shows the ideal and practical frequency responses of the notch filter. Ideally, we would like to completely block all frequencies in a particular range. Practically, we can achieve a valley or notch in the magnitude of the frequency response characteristic.

Notch filters are commonly used when a particular frequency that appears in the system needs to be removed before further processing. As an example, a system might have triplen harmonics injected into it causing the voltages to have triplen harmonics. A triplen harmonic is either a 3rd harmonic or a multiple of it—6th, 9th, etc. Quite often the 3rd harmonic might have a magnitude that is large enough to be bothersome and might disrupt the performance of controllers. Therefore, a notch filter might be an ideal candidate to remove this particular harmonic component and let everything else pass through. The question could be asked—why not block the 3rd harmonic and all other higher harmonics as well with a low pass filter? As stated before, a filter needs to be chosen with respect to the application and if the application requires for any reason only a particular frequency to be removed, the filter needs to comply with that.

Let us define the objectives of this notch filter. The notch filter is expected to attenuate a 550 Hz harmonic component in the input, while allowing the fundamental and higher order frequency components to pass through. To set a cut-off for the higher order frequency component that must pass through the filter, let us suppose that there is a higher order harmonic component of frequency 2755 Hz present in the input that needs to be passed through and should be present in the output with the minimum possible distortion. Since this is a notch filter, the filter would also be expected to pass through frequency components less than the fundamental 50 Hz component as well as frequency components greater than the 2755 Hz frequency component. With these specifications, let us begin our filter design exercise.

Since one objective of the our filter is the same as before—which is needing to block the 550 Hz frequency component, we could begin our design with a second order pole as before. In the example of the low pass filter, we had two second order poles cascaded together to more effectively remove the 550 Hz frequency component. However, we had arrived at that design iteratively. Let us repeat the process here as well and start with a single second order pole

$$H(s) = \frac{\omega_0^2}{s^2 + 2\zeta\omega_0 + \omega_0^2} \tag{7.10}$$

Let us also choose the same parameters as before—$\omega_0 = 2\pi \times 275$ rad/s, $\zeta = 0.3$. The frequency response characteristics will be the same as Fig. 7.3.

With this very first step, we can now assess the requirements going forward. The magnitude of the Bode plot continuously decreases after the resonant frequency of 275 Hz at the rate of 40 dB per decade. The magnitude of the transfer function must decrease sufficiently in order for the magnitude at 550 Hz to be low enough for that frequency component to be attenuated. However, the magnitude of the transfer function must recover to an absolute value of unity before the cut-off frequency of 2755 Hz. With reference to Fig. 7.7, the valley of the notch must be as close as possible to 550 Hz to achieve the lowest possible attenuation. With this requirement, we will examine how we can add generalized second order zeros in the next section.

7.8   Adding Zeros to the Transfer Function

In the previous section, we had started off the notch filter design with a second order pole. Our next requirement was to be able to create a valley in the frequency response characteristics. In this section, we will examine how we can use a generalized second order zero for this purpose. The transfer function of the generalized second order zero has been provided in Chap. 6 along with its Bode plots.

The transfer function of the generalized second order zero is

$$H(s) = K\frac{s^2 + 2\zeta\omega_0 + \omega_0^2}{\omega_0^2} \tag{7.11}$$

Just as with the case of the second order pole, we do not need any blanket amplification or attenuation and, therefore, the gain K can be unity or 1. As was the case with the second order pole, the magnitude of the second order zero has a resonance peak at the resonance frequency ω_0. However, in the case of a zero, the resonance peak is the inverse of a pole—during resonance, the magnitude of the zero decreases sharply. In the case of the second order pole, the resonant peak was a cause of concern as it could potentially amplify a signal of frequency close to the resonant frequency. In the case of a second order zero, the resonant peak is not a

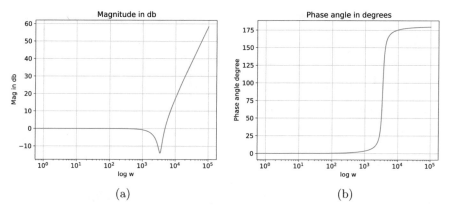

Fig. 7.8 Bode plots of the second order zero. (**a**) Magnitude plot. (**b**) Phase angle plot

cause of concern as if the magnitude decreases at the resonant peak, that is only welcome as our objective is to create a valley with the magnitude being as low as possible to attenuate the frequency close to the valley. This makes our choice of the damping factor ζ much simpler as we do not have to look for a system close to being critically damped. We could choose a low value of ζ, and let the second order zero be underdamped. Therefore, we will choose a value of ζ equal to 0.1.

The next factor is the resonant frequency ω_0. In the case of the second order zero, the magnitude of the zero will be unity until the resonant frequency. For frequencies greater than the resonant frequency, the magnitude will increase at the rate of 40 db per decade. Since we wish to create a valley at 550 Hz, the best choice for the resonant frequency of the second order zero is 550 Hz or $\omega_0 = 2\pi \times 550$ rad/s. With these parameters, the Bode plots of the second order zero are shown in Fig. 7.8.

The next stage will be to cascade the transfer functions of the second order pole and the second order zero. Mathematically,

$$H(s) = H_1(s)H_2(s) \tag{7.12}$$

where $H_1(s)$ and $H_2(s)$ are the transfer functions of the second order pole and the second order zero, respectively. We can code this cascaded transfer function in exactly the same way we did for the low pass filter. The relevant section of the code is

```
omega1  =  2*np.pi*275.0
zeta1  =  0.3
omega2  =  2*np.pi*550.0
zeta2  =  0.1

num1_2ndpole  =  [omega1*omega1]
den1_2ndpole  =  [1, 2*zeta1*omega1,  omega1*omega1]
secondpole_h  =  signal.lti(num1_2ndpole,  den1_2ndpole)
```

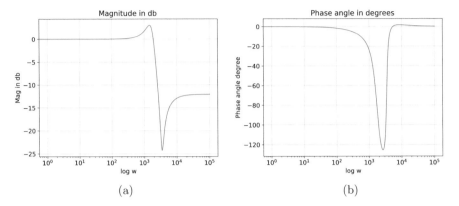

Fig. 7.9 Bode plots of the cascaded transfer function with one pole and one zero. (**a**) Magnitude plot. (**b**) Phase angle plot

```
num1_2ndzero  =  [1 ,  2*zeta2*omega2 ,  omega2*omega2]
den1_2ndzero  =  [omega2*omega2]

num_tf_h  =  np.polymul(num1_2ndpole ,  num1_2ndzero)
den_tf_h  =  np.polymul(den1_2ndpole ,  den1_2ndzero)
tf_h  =  signal.lti(num_tf_h ,  den_tf_h)
w,  mag,  phase  =  signal.bode(tf_h ,  np.arange(100000))
```

The rest of the code has been skipped to avoid repetition. However, we can use the same plotting commands as in the case of the low pass filter to generate the Bode plots for the cascaded transfer function as shown in Fig. 7.9.

At first glance, it may appear from Fig. 7.9 that a notch filter has been achieved. However, this is still not the solution we want. The first objective of passing through the fundamental 50 Hz frequency without amplification or attenuation has been maintained since the magnitude of the cascaded transfer function at 50 Hz is 0.16 dB, which is approximately equal to an absolute magnitude of 1. The phase angle of the cascaded transfer function at 50 Hz is $-5.5°$, which is also a negligible phase angle delay. The next objective is sufficiently attenuating the 550 Hz frequency component. The magnitude of the cascaded transfer function at 550 Hz is approximately -24 dB, which corresponds to an absolute magnitude of 0.063. This implies that the 550 Hz component in the output will be attenuated to 6.3% of its value in the input. This is sufficient attenuation.

The magnitude of the Bode plot is seen to increase after the resonant zero at 550 Hz, and the shape of the magnitude plot is indeed a valley. However, it is to be noted that the large dip of the magnitude at 550 Hz is due to the resonance of the second order zero. It is this resonance that gives the perception of a notch. As the frequency increases beyond 550 Hz, the magnitude of the cascaded transfer function stabilizes at approximately -12 dB, which corresponds to an absolute magnitude of 0.25. Due to this, our final objective of passing through the 2755 Hz will fail as this frequency component will be attenuated to 25% of its value. Therefore, though we have succeeded on several counts, we have not yet achieved our final notch filter design.

To understand what we should do next, let us understand the effect of cascading a pole and a zero. A second order pole results in the magnitude of the transfer function decreasing at the rate of 40 dB per decade for frequencies greater than the resonant frequency of 275 Hz. A second order zero, on the other hand, results in the magnitude of the transfer function increasing at the rate of 40 dB per decade for frequencies greater than the resonant frequency of 550 Hz. Therefore, by cascading a second order pole and a second order zero, for frequencies greater than the second resonant frequency of 550 Hz, the magnitude of the transfer function remains constant. However, in order to achieve a notch filter, the magnitude of the transfer function must revert to 0 dB and then remain constant at 0 dB. From Fig. 7.9, the magnitude of the transfer function has levelled off prematurely.

To ensure that the magnitude of the cascaded transfer function reverts back to 0 dB, we need to add another second order zero, which will continue to force the magnitude of the transfer function to increase. Since all we need is the magnitude rising for frequencies greater than 550 Hz, we can add another identical second order zero to the cascaded transfer function

$$H(s) = H_1(s)H_2(s)H_3(s) \qquad (7.13)$$

where $H_3(s)$ will be the second order zero identical to $H_2(s)$. The code block is as follows:

```
omega1 = 2*np.pi*275.0
zeta1 = 0.3
omega2 = 2*np.pi*550.0
zeta2 = 0.1
omega3 = 2*np.pi*550.0
zeta3 = 0.1


num1_2ndpole = [omega1*omega1]
den1_2ndpole = [1, 2*zeta1*omega1, omega1*omega1]
num1_2ndzero = [1, 2*zeta2*omega2, omega2*omega2]
den1_2ndzero = [omega2*omega2]
num2_2ndzero = [1, 2*zeta3*omega3, omega3*omega3]
den2_2ndzero = [omega3*omega2]


num_tf_h = np.polymul(num1_2ndpole, num1_2ndzero)
num_tf_h = np.polymul(num_tf_h, num2_2ndzero)
den_tf_h = np.polymul(den1_2ndpole, den1_2ndzero)
den_tf_h = np.polymul(den_tf_h, den2_2ndzero)
tf_h = signal.lti(num_tf_h, den_tf_h)
w, mag, phase = signal.bode(tf_h, np.arange(100000))
```

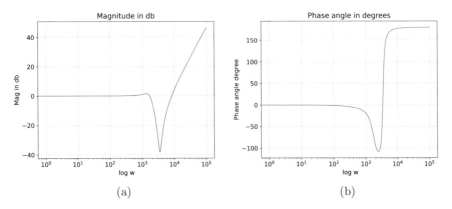

Fig. 7.10 Bode plots of the cascaded transfer function with one pole and two zeros. (**a**) Magnitude plot. (**b**) Phase angle plot

The Bode plot for this cascaded transfer function is shown in Fig. 7.10.

A quick glance at Fig. 7.10 makes one think that we have made it worse. The magnitude of the transfer function now is increasing continuously at the rate of 40 db per decade for frequencies greater than the resonant frequency. In reality, we are one step closer to the final solution as now the magnitude is rising again towards 0 dB. However, due to the double second order zero at 550 Hz, there is now a net positive rate of change of the magnitude. In the next section, we will examine how this continuous increase can be arrested.

7.9   Normalizing the Transfer Function

In the previous section, we had come one step closer to designing the notch filter by cascading a single second order pole and two second order zeros. We had seen that the drawback was that the magnitude was continuously increasing for frequencies greater than 550 Hz. In this section, we will examine how to level off the magnitude at unity so as to ensure that higher order harmonics are not amplified but are merely passed through.

We will follow a similar approach to the previous section. Since now the rate of increase of the magnitude of the transfer function is 40 dB per decade, we need to add a second order pole that will nullify the increase in magnitude. As always, the design of the second order pole will need us to choose the resonant frequency ω and the damping factor ζ. Since we are designing a second order pole, we would like to ensure that the resonant peak is limited. Therefore, as with the very first second order pole, let us choose $\zeta = 0.3$ to limit the resonant peak. The resonant frequency on the other hand is more tricky.

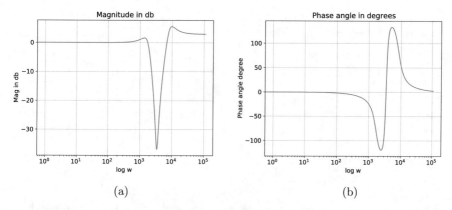

Fig. 7.11 Bode plots of the transfer function $H(s)$ with second zero at 1306 Hz. (**a**) Magnitude plot. (**b**) Phase angle plot

Since our objective behind adding a second order pole is to level off the magnitude of the transfer function, the resonant frequency of the second order pole needs to be placed at the frequency where the magnitude of the transfer function intersects the 0 dB axis. Since we have already generated the Bode plots, all we have to do is zoom in on the magnitude plot to determine the frequency at which the magnitude plot intersects the 0 dB axis. This frequency is approximately 8206 rad/s or 1306 Hz. The frequency response is shown in Fig. 7.11.

From Fig. 7.11, it is clear that this is also not the solution as the magnitude of the transfer function is not 0 dB when the magnitude levels off. A non-zero magnitude implies that the frequency components are amplified.

It is possible to arrive at a rigorous mathematical solution that avoids errors such as these. However, since we now have very convenient computational tools with which we can generate the Bode plots for higher order transfer functions, a trial and error method is far more convenient. By trial and error, it can be found that placing the second order pole with a resonant frequency of 1100 Hz levels off the magnitude of the transfer function as desired. Figure 7.12 shows the Bode plots, and it can be seen from the magnitude plot that the magnitude of the transfer function settles to 0 dB at steady state.

Let us assess in detail the performance of the transfer function for higher frequencies. Upon zooming in on the Bode plots, we find that for frequencies greater than 1910 Hz, the magnitude of the transfer function is 1 dB, which is a negligible amplification. As for the phase angle plot, we find that for frequencies greater than 1910 Hz, the phase angle of the transfer function is merely 15°, which is an acceptable phase angle advance. Therefore, the transfer function would pass through frequency components greater than 1910 Hz without significant distortion. Before we implement this cascaded transfer function, we need to address another limitation of the lti function and the associated to_discrete method.

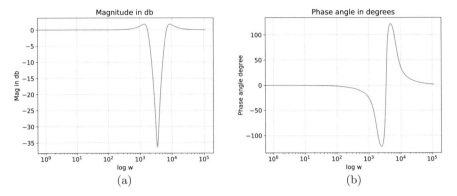

Fig. 7.12 Bode plots of the transfer function $H(s)$ with second zero at 1100 Hz. (**a**) Magnitude plot. (**b**) Phase angle plot

In the past example, when we have implemented second order poles, we used the lti function to create the transfer function object, and subsequently, we used the to_discrete() method of the transfer function object to convert the transfer function into the digital domain. When implementing a zero, this direct use of the to_discrete method is not possible. For example, if we have this code block,

```
num1_2ndzero = [1, 2*zeta2*omega2, omega2*omega2]
den1_2ndzero = [omega2*omega2]
secondzero_h1 = signal.lti(num1_2ndzero, den1_2ndzero)
secondzero_h1_z = secondzero_h1.to_discrete(
        dt=200.0e-6,
        method='tustin'
        )
```

We will get the following error:

```
ValueError: Improper transfer function. 'num' is
   longer than 'den'.
```

An improper transfer function is when the order of the numerator polynomial is greater than the order of the denominator polynomial. This is the case with the second order zero or for that matter any zero. The error that we encounter is an error with the to_discrete method and is not a mathematical limitation. It is perfectly legal to convert a second order zero transfer function from the continuous s domain into the digital z domain. Therefore, we need to find a work around for this limitation of the to_discrete method. The solution is to represent the second order zero as a pole instead and while implementing it, use the reciprocal instead. This would be the code block instead

```
num1_2ndzero = [1, 2*zeta2*omega2, omega2*omega2]
den1_2ndzero = [omega2*omega2]
secondzero_h1 = signal.lti(den1_2ndzero, num1_2ndzero)
```

```
secondzero_h1_z = secondzero_h1.to_discrete(
    dt=200.0e-6,
    method='tustin'
    )

// — other code
    u2[0] = y1[0]
    y2[0] = ( secondzero_h1_z.den[0] * u2[0] + \
        secondzero_h1_z.den[1] * u2[1] + \
        secondzero_h1_z.den[2] * u2[2] - \
        secondzero_h1_z.num[1] * y2[1] - \
        secondzero_h1_z.num[2] * y2[2] ) / 
        secondzero_h1_z.num[0]
    u2[2] = u2[1]
    y2[2] = y2[1]
    u2[1] = u2[0]
    y2[1] = y2[0]
```

It can be observed that the second order zero has been synthesized using the reciprocal to mimic a second order pole. During implementation, the numerator and denominator of the digital transfer function have been reversed to achieve the reciprocal which is the original second order zero. This is a convenient trick to implement improper transfer functions using Python.

The entire code for the cascaded filter is as follows:

```
import numpy as np
import matplotlib.pyplot as plt
from scipy import signal

omega1 = 2*np.pi*275.0
zeta1 = 0.3
omega2 = 2*np.pi*550.0
zeta2 = 0.3
omega3 = 2*np.pi*550.0
zeta3 = 0.1
omega4 = 2*np.pi*1100.0
zeta4 = 0.3

num1_2ndpole = [omega1*omega1]
den1_2ndpole = [1, 2*zeta1*omega1, omega1*omega1]
secondpole_h1 = signal.lti(num1_2ndpole, den1_2ndpole)
num1_2ndzero = [1, 2*zeta2*omega2, omega2*omega2]
den1_2ndzero = [omega2*omega2]
secondzero_h1 = signal.lti(den1_2ndzero, num1_2ndzero)
num2_2ndzero = [1, 2*zeta3*omega3, omega3*omega3]
den2_2ndzero = [omega3*omega2]
```

```python
secondzero_h2 = signal.lti(den2_2ndzero, num2_2ndzero)
num2_2ndpole = [omega4*omega4]
den2_2ndpole = [1, 2*zeta4*omega4, omega4*omega4]
secondpole_h2 = signal.lti(num2_2ndpole, den2_2ndpole)
secondpole_h1_z = secondpole_h1.to_discrete(
        dt=200.0e-6,
        method='tustin'
        )
secondzero_h1_z = secondzero_h1.to_discrete(
        dt=200.0e-6,
        method='tustin'
        )
secondzero_h2_z = secondzero_h2.to_discrete(
        dt=200.0e-6,
        method='tustin'
        )
secondpole_h2_z = secondpole_h2.to_discrete(
        dt=200.0e-6,
        method='tustin'
        )

t_duration = 1.0
t_step = 1.0e-6
no_of_data = int(t_duration/t_step)
time_array = np.arange(no_of_data)*t_step
frequency = 50.0
omega = 2*np.pi*frequency
omega_noise = 2*np.pi*550.0
omega_hf = 2*np.pi*2755.0
inp_mag = np.sqrt(2)*240.0
ip_voltage_signal = inp_mag*(
                np.sin(time_array*omega) + \
                0.0*np.sin(time_array*omega_noise) + \
                0.1*np.sin(time_array*omega_hf)
            )
t_sample = 200.0e-6
no_of_skip = int(t_sample/t_step)
tsample_array = time_array[::no_of_skip]
ip_voltage_samples = ip_voltage_signal[::no_of_skip]
op_voltage_samples = np.zeros(ip_voltage_samples.size)

u1 = np.zeros(3)
y1 = np.zeros(3)
u2 = np.zeros(3)
```

```python
y2 = np.zeros(3)
u3 = np.zeros(3)
y3 = np.zeros(3)
u4 = np.zeros(3)
y4 = np.zeros(3)

for volt_index, volt_value in np.ndenumerate
(ip_voltage_samples):
    u1[0] = volt_value
    y1[0] = ( secondpole_h1_z.num[0] * u1[0] +\
        secondpole_h1_z.num[1] * u1[1] + \
        secondpole_h1_z.num[2] * u1[2] - \
        secondpole_h1_z.den[1] * y1[1] - \
        secondpole_h1_z.den[2] * y1[2] ) / \
        secondpole_h1_z.den[0]
    u1[2] = u1[1]
    y1[2] = y1[1]
    u1[1] = u1[0]
    y1[1] = y1[0]

    u2[0] = y1[0]
    y2[0] = ( secondzero_h1_z.den[0] * u2[0] + \
        secondzero_h1_z.den[1] * u2[1] + \
        secondzero_h1_z.den[2] * u2[2] - \
        secondzero_h1_z.num[1] * y2[1] - \
        secondzero_h1_z.num[2] * y2[2] ) / \
        secondzero_h1_z.num[0]
    u2[2] = u2[1]
    y2[2] = y2[1]
    u2[1] = u2[0]
    y2[1] = y2[0]

    u3[0] = y2[0]
    y3[0] = ( secondzero_h2_z.den[0] * u3[0] + \
        secondzero_h2_z.den[1] * u3[1] + \
        secondzero_h2_z.den[2] * u3[2] - \
        secondzero_h2_z.num[1] * y3[1] - \
        secondzero_h2_z.num[2] * y3[2] ) / \
        secondzero_h2_z.num[0]
    u3[2] = u3[1]
    y3[2] = y3[1]
    u3[1] = u3[0]
    y3[1] = y3[0]
```

```
    u4[0]  =  y3[0]
    y4[0]  =  (  secondpole_h2_z.num[0]  *  u4[0]  +  \
       secondpole_h2_z.num[1]  *  u4[1]  +  \
       secondpole_h2_z.num[2]  *  u4[2]  -  \
       secondpole_h2_z.den[1]  *  y4[1]  -  \
       secondpole_h2_z.den[2]  *  y4[2]  )  /
       secondpole_h2_z.den[0]
    u4[2]  =  u4[1]
    y4[2]  =  y4[1]
    u4[1]  =  u4[0]
    y4[1]  =  y4[0]

    op_voltage_samples[volt_index]  =  y4[0]

plt.figure()
plt.plot(tsample_array,  ip_voltage_samples,  \
                         label='input',  ds='steps')
plt.plot(tsample_array,  op_voltage_samples,  \
                         label='output',  ds='steps')
plt.title('Input  versus  output')
plt.xlabel('Time')
plt.ylabel('Volts')
plt.grid()
plt.legend()
plt.show()
```

As can be seen from the code block above, the ip_voltage_signal has a 30% 550 Hz harmonic content and a 10% 2755 Hz frequency component. In order to fully understand the performance of the transfer function as a notch filter, we will examine three different cases of the ip_voltage_signal. As a first case, let us examine the output of the filter when ip_voltage_signal contains only the fundamental 50 Hz frequency component and the 2755 Hz frequency component. The result is shown in Fig. 7.13, where a single cycle of the fundamental is zoomed in. As can be seen, the output follows the fundamental 50 Hz frequency component fairly accurately and the 2755 Hz frequency component with a small phase angle delay. Therefore, the notch filter passes through low frequencies and high frequencies with minimal distortion.

To examine the performance of the notch filter in blocking the 550 Hz frequency component, the fundamental 50 Hz frequency component and the 2755 Hz frequency component are set to zero in ip_voltage_signal, and the 550 Hz component is preserved as 30% of the rated voltage magnitude. As can be seen from Fig. 7.14, the output signal is around 7 Volts, while the input is around 100 Volts. This implies that the 550 Hz frequency component has been attenuated effectively.

Finally, the ip_voltage_signal contains all three frequency components—the fundamental 50 Hz, the 550 Hz and the 2755 Hz. The performance of the filter is

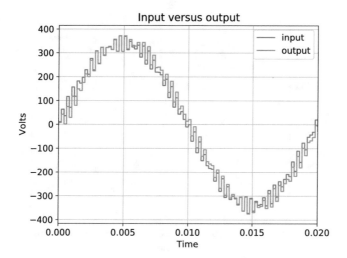

Fig. 7.13 Performance of the notch filter—passing through components

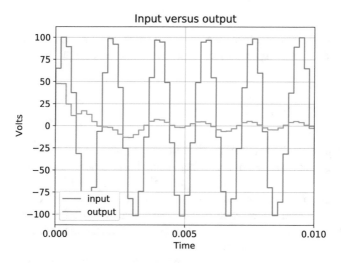

Fig. 7.14 Performance of the notch filter—blocking components

shown in Fig. 7.15. As can be seen, the filter preserves the low and high frequency components, while blocking the 550 Hz frequency component.

In this section, we completed the design of the notch filter by cascading four lower order transfer functions and fine-tuning the parameters until the frequency response characteristics fulfilled our requirements. Using simulations, we verified the performance of the notch filter for the specifications that we had set out with while designing the filter. With this, we have demonstrated how fairly complex filters can be designed by using lower order generalized transfer functions as

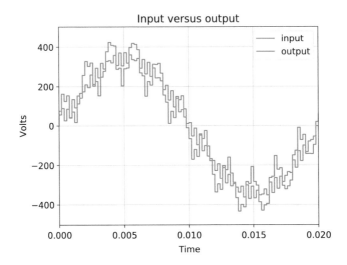

Fig. 7.15 Performance of the notch filter

building blocks. Though there are advanced filters available in the literature, this procedure allows a designer to use simple techniques to design filters.

7.10 Conclusions

In this chapter, we have put to practice the theory that we have learned in the past few chapters. We have synthesized filters using basic generalized transfer functions whose behaviour is well known and presented in the previous chapter. We have started our design examples by laying down the specifications of the filter and gradually cascading generalized transfer functions. The design process has been incremental with analysis of frequency response characteristics and the verification of performance through simulations.

This chapter has used the minimal of manual calculations and has used only computational tools for the design and verification process. As a filter designer, this process is fast, convenient and has a low chance of error as it requires only parameters to be specified with all other computations performed by programming. We have introduced several new commands in Scipy as well as NumPy to be able to design and verify filters. We have examined how the to_discrete method available with TransferFunction objects created by the lti function can be used to convert a transfer function from the continuous s domain into the digital z domain. We have also examined how the polymul function available with NumPy can be used to synthesize higher order transfer functions by cascading lower order transfer functions.

With examples of a low pass filter and a notch filter, this chapter serves as reference for any engineer who wishes to design filters for any particular objective. The method followed in this chapter can be replicated for any signal processing assignment. Each filter example consists of a specification stage that leads us to a first hypothesis. Frequency response analysis and simulation are used to assess the hypothesis and provide us insights to improve the performance. With each improvement, analysis and simulation are repeated until we arrive at the desired filter.

References

1. Dzhankhotov, V., Pyrhnen, J., Silventoinen, P., & Kuisma, M. (2012). Hybrid lc filter electrical design considerations. *IEEE Transactions on Industrial Electronics, 59*(02), 762–768.
2. Zverev, A. I. (2005). *Handbook of Filter Synthesis* (Vol. 1). Wiley.
3. Williams, A. B. (1995). *Electronic Filter Design Handbook*. McGraw-Hill.

Chapter 8
Conclusions

8.1 A Summary of the Course Contents

The motivation behind writing this book was to introduce signal processing not as a subject but as a set of tools that an engineer can use. The objective was to design filters—a task which most power engineers find themselves doing regularly. The aim of this book was to use fundamentals from signal processing to create a well-structured method for designing filters. Signal processing theory can be used for various purposes, and the interested reader can always find other resources to expand on this knowledge. However, my aim behind writing this book was to describe theory with a concrete objective in mind.

In this book, we described the basics of digital systems with a few engineering and non-engineering examples. We described how digital systems are conceived and can be implemented using easily available hardware. We then progressed to signal processing theory. However, we eased into signal processing theory using examples from analog circuits to make the process easier for a power engineer to understand. We introduced Laplace Transform and showed how it converts a time domain equation to a frequency domain equation. We examined the advantages of using Laplace Transform and how it results in equations that are easier to analyze.

To make it clear how the Laplace Transform converts a time domain equation to a frequency domain equation, we meandered through a few mathematical concepts. We compared how frequency is defined in the power engineering domain and how it is defined in the mathematical domain. We examined how a sinusoid is expressed in power engineering versus how it is expressed in mathematics. We then examined how the Laplace Transform uses the mathematical expression for sinusoids to introduce frequency into the transformed equation. Though every signal processing book deals with Laplace Transform in great detail, this basic description of how frequency is introduced into the transformed equation is often missing.

© The Editor(s) (if applicable) and The Author(s), under exclusive license to
Springer Nature Switzerland AG 2020
S. V. Iyer, *Digital Filter Design using Python for Power Engineering Applications*,
https://doi.org/10.1007/978-3-030-61860-5_8

After the Laplace Transform, we examined how the transformed equation is now in the frequency domain but is still continuous. We examined how the frequency can also be perceived as an operator with a comparison to analog circuits. We then introduced the process of conversion of a continuous frequency system to a digital or discrete frequency system. We examined a few commonly used methods of conversion and presented the Bilinear Transformation as the most stable conversion process.

After converting an equation in the continuous frequency domain to the digital frequency domain, the next step is to implement this equation. We again project the digital frequency z as an operator in a manner similar to how the continuous frequency s was an operator. With the digital frequency z also defined as an operator, we then converted the digital equation in z to a difference equation in the time domain. This difference equation in the time domain is expressed with respect to the samples of the input and output signals in the equation.

To describe how this entire process of converting a continuous time equation to a difference equation can be used to implement a filter, we take up the examples of the analog LC filter. The LC filter is one of the most commonly used low pass filters in power engineering. We expressed the differential time domain equations for the LC filter and using Laplace Transform converted them to continuous frequency domain equations. We then used Bilinear Transformation to convert the continuous frequency equations in s to the digital frequency equations in z. With the definition of z as an operator, we wrote the difference equation for the LC filter in terms of discrete time samples. We simulated the difference equation using Python and were able to examine the performance of the LC filter by plotting the output signal with respect to the input signal.

Now that we simulated a digital filter and established a way to verify its performance, the next step was to design a filter. We then examined frequency response characteristics of a transfer function and learned how frequency domain equations can be analyzed for their behaviour at different frequencies. We introduced the concept of Bode plots. We revisited the LC filter and generated its Bode plot to analyze its performance. We then introduced generalized transfer functions that have well-defined formats that result in predictable frequency response characteristics. We analyzed the Bode plots for the generalized first order pole, the second order pole, the first order zero and the second order zero.

To bring together all these concepts, we chose to design a low pass filter and a band stop (notch) filter. We used the generalized transfer functions as building blocks to synthesize our filters. We used an incremental step-by-step approach—design using generalized transfer functions, analysis using frequency response characteristics and verification using simulation of the digital implementation of the filter. We used Python commands to automate all these tasks with the only manual input being the specification of the parameters of the transfer functions.

8.2   A Few Comments on the Approach

The approach to teaching signal processing in this book may seem narrow and too focused. Most signal processing books cover vast amount of theory. Signal processing is an advanced topic, and for anyone to consider themselves to be well read would imply reading over a dozen books. In comparison, this book talks only about filter design. This book may seem completely useless in comparison to regular reference books on signal processing. For this reason, it is important that I spend some time talking about why I chose this approach.

As stated in the introduction, the motivation for writing this book was to introduce signal processing to power engineers. The book instead of describing theory with some examples and exercises uses the reverse approach—it describes theory so as to be able to achieve a particular task. The task chosen has been designing digital filters which is a task quite often undertaken by a power engineer. This book therefore has a fixed purpose. And this book takes the approach—the end justifies the means. Therefore, what is needed to be able to design digital filters?

It is important to understand how a filter behaves as a filter. How does a filter block a particular set of frequencies and let other pass through? To understand this, the frequency response characteristics need to be understood with the computational tools to be able to generate frequency response characteristics. To be able to understand how a system varies with frequency, we need to express a system in the frequency domain. Conventionally, our expressions are in the time domain. Therefore, how do we start in the time domain and move to the frequency domain?

To achieve this, we introduce the Laplace Transform and apply it to the time domain equations of well-known filters such as the LC filter. We examine the frequency response characteristics of the transfer function of the LC filter and infer the behaviour of the LC filter with the frequency response characteristics. With this, we figured out how frequency response characteristics can be used to understand how a system behaves with changing frequency. This laid the foundation for designing a filter.

Besides designing a filter, it is also essential to know how to implement the filter. To understand how this can be done, we have shown how a continuous transfer function can be converted to a digital transfer function. Subsequently, we used the digital transfer function to express the final system as a difference equation. To verify our filter designs, we simulated the difference equation of the filter. However, the implementation in the simulation will be the same as the implementation in a hardware platform such as a micro-controller.

This book is therefore rather a manual than a reference book. The focus of the book is to enable practicing engineers to be able to design filters using a process that is automated and repeatable. The theory in the book is to be able to appreciate the final process and understand why each step counts.

8.3   Scope for Future Work

For anyone who has read this book and has reached until this section, I hope that you have been coding along. The main objective behind this book has been to provide practicing engineers the tools needed to be able to design filters. In addition, this book uses only free and open source software—Python, NumPy, SciPy and MatPlotLib. Therefore, the tools are available to you for free and without any licensing restrictions.

If you, as the reader, have been able to implement the low pass and notch filters that have been covered in the book, the next step is to use this process and try to use it for your project. The implementation of digital filters as difference equations needs only a few samples of the input and output. Therefore, filters designed using the process outlined in this book can be implemented in any simulation software.

In this book, the filter design was achieved using simple low order transfer functions. However, there is a vast array of very advanced higher order filters that can also be used. An interested reader is welcome to use the same techniques to implement higher order filters and examine the difference in the performance.

As an online teacher and author, I have made it my goal to promote the use of open source software in engineering. In this book, I have used Python and associated tools to design filters. The Python functions that we have used are merely the tip of the iceberg. Python has a vast array of packages for a number of different purposes. Many of these packages are being aggressively developed by groups of developers. It is strongly encouraged that the reader tries to adopt some of the open source packages as this will lead to their growth and popularity. Eventually, the goal is to make education universally accessible to every interested person on this planet irrespective of their geographic location, financial situation or time constraint.

Index

A

ADC, *see* Analog to digital converter (ADC)

Analog filter
  branch with capacitor, 72
  capacitor, 72–74
  coding
    capacitor, 74–82
    inductor, 84–88
  dc offset, 88–90
  digital filtering, 5
  inductor filter, 82–84
  Laplace Transform, 71
  LC filter modeling, 102–106
  lossy
    capacitor, 92–98
    inductor, 98–102
  offsets in reality, 90–92
  performance, LC filter, 106–110

Analog to digital converter (ADC)
  comparators, 19
  process of quantization, 18
  processor, 19–20
  sample and hold circuit, 18

Arrays
  of angles, 60
  functions, 4, 56
  initializations, 78
  magnitude, 130
  manipulation, 57–58
  Numpy, 55–57
  packages, 54
  plotting, 62, 63

Array slicing, 63, 69, 76, 84

B

Bilinear Transformation, 49, 52, 72, 83, 103, 134, 149, 157, 158, 188
  continuous
    domain frequency expression, 83
    frequency equations, 72, 188
  conversion process, 167
  Euler's method, 158
  transfer function, 149, 158
  Trapezoidal Rule, 49

Bode plots, 121–124
  cascaded transfer function, 174
  code, 131
  frequency argument, 129
  inductor transfer function, 131
  logarithm, 133
  Matplotlib, 130
  phase angle, 140
  powerful tool, 113
  second order zero transfer function, 156
  signal package, 129
  transfer function, 149

C

Continuity and differentiability, 47
Continuous time signal processing, 11–12

D

DC offset, 88–90
  ac wave, 90
  equation, 92

© The Editor(s) (if applicable) and The Author(s), under exclusive license to
Springer Nature Switzerland AG 2020
S. V. Iyer, *Digital Filter Design using Python for Power Engineering Applications*,
https://doi.org/10.1007/978-3-030-61860-5

DC offset (*cont.*)
    integration, 92
    loss dissipation, 97
    output current, 98
    physical significance, 93
Digital implementation, 2, 3, 5, 7, 76, 138,
            157, 160, 161, 170, 188
Digital signal processing (DSP)
    advantages of, 15–16
    circuits, 13
    computational tools, 4–5
    digital technologies, 1
    filter, 3, 14
    FPGAs, 3
    implementation, 2
    need, 12–13
    outline, 6–8
    philosophy, 5–6
    power system, 1
    processor, 14
    See also Signal processing
Digitization of power systems, 3
Discrete systems
    ADC, 19–20
    vs. continuous, 10–11
    DSP, 12–15
    practical analog to digital conversion,
            18–19
    signal to digital form convertion, 16–18
    time signal processing, continuous, 11–12
DSP, *see* Digital signal processing (DSP)

E

Euler's equation, 29, 30, 38
    Laplace Transform, 39
    sine/cosine, 37
    sinusoids, 41
    vector's projection, 38

F

Field programmable gate arrays (FPGAs), 3,
            13
Filter behaviour interpretation, 113, 189
Filter design
    cascaded low pass, 166–171
    low pass, 151–157
    normalizing the transfer function, 176–184
    notch filter, 170–172
    NumPy, 149
    power applications, 150–151
    SciPy, 149
    simulation, 157–162

    transfer functions cascading, 157–166
    zeros addition, 172–176
FPGAs, *see* Field programmable gate arrays
            (FPGAs)
Frequency response
    bode plots, 121–124
    complex numbers, 114–115
    inductor, 127–130
    interpreting the inductor, 130–134
    LC filter, 113, 134–138
    physical significance, 138–140
    poles
        first order, 140–143
        second order, 143–144
    transfer functions (*see* Transfer functions)
    zero
        first order, 144–145
        second order, 145–147

G

Grid frequency, 138, 139, 151, 156

I

Incremental filter design, 5, 149
Interrupt service routine (ISR), 82

L

Laplace Transform, 28–31
    advantage of, 41–43
    continuous frequency domain equations,
            188
    digital domain, 45, 74
    frequency domain representation, 7
    inverse, 35
    LC filter, 189
    mathematical expression, 187
    negative sign, 39
    passive components, 6
    polynomials, 50
    revisiting inductors and capacitors, 31–34
    time varying
        equations, 52
        function, 44
LC filter, 25, 27, 52, 71
    digital implementation, 161
    frequency response, 134–138
    Laplace Transform, 189
    modeling, 102–105
    in power engineering, 188
    simulation, 111
Linear time invariant (LTI) systems, 125

Low pass filters
    band stop/notch filter, 149
    cascaded, 166–170
    design, 151–157
    filter
        LC, 52
        pass, 151
    frequency components, 150
LTI systems, *see* Linear time invariant (LTI)
        systems

M
Matplotlib
    library plotting, 5
    open source software, 190
    packages, 55
    plotting, 60–64
Microcontrollers
    circuit breaker, 2
    DSP, 14
    filter, 167
    FPGAs, 3, 13
    ISR, 82
    and microprocessors, 16
    power electronics application, 3
    price of, 2
    second order pole, 157
Microprocessors, 13, 16

N
Notch filter, 170–172
    digital filtering, 13
    frequency range, 151
    passive components, 12
    performance, 183, 184
    second order zeros, 176
Numeric Python (Numpy)
    array
        of functions, 4
        slicing, 76
    digital filter input, 85
    and Matplotlib, 53, 75
    polynomials, 164
    scientific computation, 59
    signals generation, 55–57
Numpy, *see* Numeric Python (Numpy)

O
Open source software, 4, 54, 190

P
Passive components, 6, 12, 15, 23, 42, 52,
        143
Plotting
    commands, 153
    library, 5
    Matplotlib, 60–64
    output signal, 188
    waveforms, 64–69
Polynomial equations, 27, 34, 42
Power system, 1–3
Python
    array manipulation, 57–58
    generating signals, 55–57
    Matplotlib, 53
    NumPy, 53
    plotting
        with Matplotlib, 60–64
        waveforms, 64–69
    programming setup, 54–55
    waveforms, 58–60

Q
Quantization, 6, 9, 17–19

S
Sampling
    ADC, 19
    digital filter, 58
    DSP, 20
    frequency, 45
    and quantization, 6, 9, 18
    time, 159, 167
Scientific Python (SciPy), 4, 5, 54, 69,
        124–127, 134, 149–152, 184, 190
SciPy, *see* Scientific Python (SciPy)
Signal processing
    capacitors, 23–27
    continuous to digital conversion, 48–51
    DSP, 43–48
    filters, 25–27
    inductors, 23–27
    Laplace transform, 28–31
    original variables, 34–35
    revisiting inductors and capacitors, 31–34
    transformations, 27–28
    variable s, 35–41
Significance of approach, 23, 35, 39, 41, 50,
        92, 114, 115, 130, 136, 138–140,
        149

Simulations
    dc offsets, 90
    digital
        domain, 5
        implementation, 188
    hardware, 157
    LC filter, 106
    performance, 149
    programmatic, 7
    Python code, 71

T
Transfer functions
    bode plots, 131, 144
    cascading, 162–166
    damping factor, 155

frequency response, 119–121
higher order, 170
LC filter, 139, 189
magnitude, 115–119, 123, 132
manual calculations, 149
normalizing, 176–184
phase angle, 115–119
pre-multiplier, 126
Python commands, 7
simulation, 157–162
zeros addition, 172–176
Transformations
    bilinear (*see* Bilinear Transformation)
    concept, 27–28
    integral operation, 41
    Laplace (*see* Laplace Transform)
    Trapezoidal rule, 49